Recent Advances in Energy Storage Materials and Devices

by
Li Lu and Ning Hu

In the Li-ion battery, Ions' transportation between positive and negative electrodes relies on organic-based electrolyte with a low flammable point. Therefore, use of this type of electrolyte leads to a lot of safety issues. In addition, the potential window of commercially available organic electrolytes is only up to about 4.5V, resulting that those cathode materials with high potentials cannot be used. Therefore, it is important to address above issues by developing multi-electron electrode materials, explore new nonflammable electrolyte and new battery formats.

This book compiles nine contributions from the principle of Li-ion batteries, cathode and anode electrode materials to future energy storage systems such as solid electrolyte for all-solid-state batteries and high capacity redox flow battery. The chapters provide the readers with comprehensive understanding of the fundamentals of energy storage through Faradic and non-Faradic mechanisms and most importantly materials and devices for future energy storage.

Contributor to the cover page
Jani Kotakoski, Associate Professor, University of Vienna, Austria

Cover page figure caption
Schematic illustration of two scanning probe microscope tips on opposing sides of graphene.

Recent Advances in Energy Storage Materials and Devices

Edited by

Li Lu[1] and Ning Hu[2]

[1]Department of Mechanical Engineering, National University of Singapore, Singapore

[2]College of Aerospace Engineering, Chongqing University, P.R. China

Published by **Materials Research Forum LLC**
Millersville, PA 17551, USA

Published as part of the book series
Materials Research Foundations
Volume 12 (2017)
ISSN 2471-8890 (Print)
ISSN 2471-8904 (Online)

Print ISBN 978-1-945291-26-5
ePDF ISBN 978-1-945291-27-2

This book contains information obtained from authentic and highly regarded sources. Reasonable efforts have been made to publish reliable data and information, but the author and publisher cannot assume responsibility for the validity of all materials or the consequences of their use. The authors and publishers have attempted to trace the copyright holders of all material reproduced in this publication and apologize to copyright holders if permission to publish in this form has not been obtained. If any copyright material has not been acknowledged please write and let us know so we may rectify in any future reprint.

Distributed worldwide by

Materials Research Forum LLC
105 Springdale Lane
Millersville, PA 17551
USA
http://www.mrforum.com

Manufactured in the United State of America
10 9 8 7 6 5 4 3 2 1

Table of Contents

Preface

Since the first commercial Li-ion battery was introduced to the market in 1991, this form of battery has gradually dominated the energy storage market, from small energy capacity batteries for portable electronics to large format batteries for electric vehicles. The capacity level has increased from less than 1 Ah cells then to over 100 Ah cells now. The electrode materials have also been developed from simple lithium cobalt oxide to many other different types of positive electrode materials in order to cater to the different needs for various applications.

Although, the Li-ion battery has truly demonstrated its advantages in energy density, power density and durability in comparison with other types of rechargeable batteries, the Li-ion battery has also its weak points. Since all commercial positive electrode materials are designed based on single electron transfer, the capacity is limited to 145 to 170 mAh g^{-1}, resulting in low energy density of about 150 to 180 Wh kg^{-1}. In the Li-ion battery, Ions' transportation between positive and negative electrodes relies on organic-based electrolyte with a low flammable point. Therefore, use of this type of electrolyte leads to a lot of safety issues. In addition, the potential window of commercially available organic electrolytes is only up to about 4.5V, resulting that those cathode materials with high potentials cannot be used. Therefore, it is important to address above issues by developing multi-electron electrode materials and explore new nonflammable electrolyte and new battery formats. This book compiles nine contributions from the principle of Li-ion batteries, cathode and anode electrode materials to future energy storage systems such as solid electrolyte for all-solid-state batteries and high capacity redox flow battery. We truly believe that these chapters will provide you comprehensive understanding of the fundamentals of energy storage through Faradic and non-Faradic mechanisms and most importantly materials and devices for future energy storage.

Li Lu, National University of Singapore, Republic of Singapore

Ning Hu, Chongqing University, People's Republic of China

Chapter 1

Lithium Ion Batteries: Fundamentals and Progress

Lu Jia*[a], Lee Kim Seng[b]

Department of Mechanical Engineering, National University of Singapore, Singapore 117576

[a]lujia@u.nus.edu, [b]mpeleeks@u.nus.edu

Abstract

Rechargeable batteries of high energy and high power density are extremely desirable for portable electronics and especially for transportations. This chapter first introduces the working principle and recent development of lithium ion batteries. Special attention is then given to the limitations of current lithium ion batteries for high energy and high power applications. To solve this problem, an overview of cathode materials is provided, showing their potential of future application in advanced vehicles.

Keywords

Lithium Ion Batteries (LIB), Principle of LIB, Development of LIB, Energy and Power Limitation of LIB, Cathodes

Contents

1.   Introduction

Growing environmental concerns such as fossil-fuel elimination, environmental pollution and global warming have driven enormous interest in renewable energies like solar and wind. However, these energy resources are not only intermittent and uncontrollable but also require efficient electrical energy storage (EES) system to deliver power stably and consistently. On the other hand, the exponential growth of portable electronic products such as video cameras, mobile phones and computers over the past decades has attracted increasing interest in efficient EES systems. Among various EES systems, rechargeable batteries are appealing for the above applications due to their high energy density and high efficiency. Lithium ion batteries (LIBs) are the most promising ones in comparison with other conventional rechargeable batteries including Ni-Cd, nickel-metal hydride and lead-acid batteries. The outstanding properties of LIBs are as follows:

1. High operating voltage.

2. High gravimetric and volumetric energy densities.

3. No memory effect.

4. Low self-discharge rate.

5. Operation in a wide range of temperature.

Research on the LIB started in the early 1980s, and first the commercial LIB was launched in 1991 by Sony. Since then, LIBs have grown to become the dominant power storage solution for portable IT devices, and are penetrating into the market of transportations recently to replace conventional fuels and eliminate CO_2 emission. More and more electric vehicles such as hybrid electric vehicles (HEVs) and plug-in hybrid electric vehicles (PHEVs) arose. However, great challenges are faced by LIBs when compare with the conventional internal combustion engine, which mainly focus on their cycling life, energy/power density, safety, and overall cost.

This chapter first introduces the technology of the LIBs as well as several basic concepts. The development of LIBs and their recent technological trends are discussed then. After a brief overview of the status and challenges of the main components for the LIBs, the attention is focused on the state-of-the-art and future of various cathode materials for large-loading LIBs, particularly for EVs and HEVs.

2.   Working Principle of LIBs

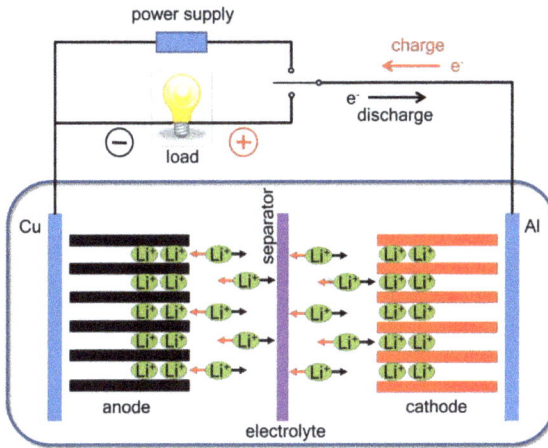

Figure 1. Schematic illustration of lithium ion batteries

LIB stores energy by converting electrical energy into electrochemical energy. The basic working principle of LIB is shown in Fig. 1. Generally, the LIB is composed of cathode, anode, electrolyte and separator in between the cathode and anode. Both cathode and anode materials are intercalation materials in the present commercial LIB systems and act as the sink for Li^+ ions while the electrolyte and separator provide a separation of ions from electronic transport. During charging, Li^+ ions are extracted from the cathode, move through the electrolyte, and are then inserted into the anode. At the same time, the electrons move from cathode to anode through the outside electronic circuit. During discharging, this process is reversed. It can be seen that the electrode materials should be both ionically and electronically conducting. In a commercial LIB where $LiCoO_2$, graphite, $LiPF_6$ in ethylene carbonate (EC)/dimethyl carbonate (DMC)/diethyl carbonate (DEC) and microporous polyethylene membrane are used as cathode, anode, electrolyte and separator, the following reactions take place during the charge-discharge process:

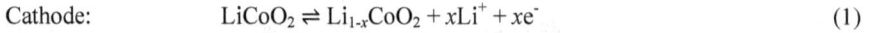

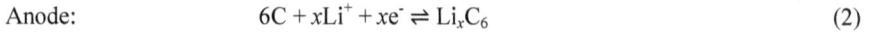

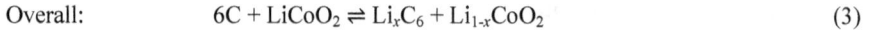

Cathode: $\text{LiCoO}_2 \rightleftharpoons \text{Li}_{1-x}\text{CoO}_2 + x\text{Li}^+ + xe^-$ (1)

Anode: $6\text{C} + x\text{Li}^+ + xe^- \rightleftharpoons \text{Li}_x\text{C}_6$ (2)

Overall: $6\text{C} + \text{LiCoO}_2 \rightleftharpoons \text{Li}_x\text{C}_6 + \text{Li}_{1-x}\text{CoO}_2$ (3)

As can be seen, reduction occurs at one side of a LIB and oxidation takes place at the other side. Overall, both reactions take place simultaneously and the combined reaction is called a Redox reaction.

In addition, some basic concepts and requirements in LIBs are necessary for a better understanding of LIBs.

2.1 Electromotive Force

The electromotive force (EMF) of a LIB, which is also called cell voltage for the external current (E_{cell}), can be identified as the difference in the standard electrode potentials of the two half Redox reactions in standard conditions:

$$E_{cell} \text{ (V)} = \varphi_{cath} - \varphi_{an}$$ (4)

where φ_{cath} and φ_{an} represent the standard electrode potentials of the cathode and anode materials (V), respectively.

In thermodynamics, the decrease in Gibbs free energy (ΔG) equals to the non-expansion work that can be extracted from a thermodynamically closed system in a completely reversible process at a constant temperature and pressure. In LIBs, the non-expansion work refers to electrical work:

$$\Delta G \text{ (J)} = -nFE_{cell}$$ (5)

where n stands for the number of moles of electrons transferred in the reaction, and F is Faraday constant (96485 C mol^{-1}). Thus,

$$E_{cell} = -\Delta G/(nF) = |(\mu_{cath} - \mu_{an})/(nF)|$$ (6)

where μ_{cath} and μ_{an} are the chemical potentials of Li ion into the cathode and anode materials. From this, the EMF of a LIB is determined by the differential chemical potential between the cathode and the anode.

So far, the E_{cell} discussed well describes the voltage when a LIB is an open circuit. In this case, the LIB seems to work in conditions close to the standard one. However, real LIBs usually work under conditions that differ from the standard conditions. The internal resistance (r) should be considered in the real LIB. During a charge with a given current I, the output voltage V_{ch} increases from the open–circuit voltage E_{cell} by an overvoltage ($\eta = I_{ch}r$). In contrast, the output voltage V_{dis} on discharge decreases by a polarization ($\eta = I_{dis}r$).

$$V_{ch} = E_{cell} + I_{ch}r \qquad (7)$$

$$V_{dis} = E_{cell} - I_{dis}r \qquad (8)$$

Therefore, the internal resistance should be as small as possible to reduce the polarization and voltage drop of a real LIB.

2.2   Internal Resistance

Section 2.1 reveals the existence of the internal resistance or impedance in a real LIB working under a given current I. According to Thévenin's theorem, a real LIB can be represented as a voltage source in series with a resistance. As discussed, when the LIB discharges, a voltage drop is introduced by the internal resistance. Besides, high resistance will also cause the LIB to heat up under cycling, which may trigger an early shutdown. Therefore, precise knowledge of the internal resistance for a LIB is of great importance for the design of specific applications. Generally, lower resistance improves the battery ability to deliver the needed power. This is especially important in heavy loads such as power tools and electric powertrains.

In practice, the internal resistance of a LIB changes along with its size, properties of electrodes, temperature, usage duration, and the current. It consists of an electronic component and an ionic component. The electronic component is mainly caused by the resistivity of the electrode materials while the ionic component is associated with electrode surface area, ion mobility, electrolyte conductivity and so on. Research laboratories have been using electrochemical impedance spectroscopy (EIS) to evaluate the internal resistance of LIBs, in which an alternating current is applied to the LIBs, typically at a frequency of 100 kHz, to estimate the resistance.

2.3   Specific Capacity

The theoretical capacity, Q_{th}, of the electrodes can be calculated according to their compositions by Faraday's law:

$$Q_{th} \text{ (mA h g}^{-1}) = \xi nF/(3600M) \tag{9}$$

where ξ is the extent of reaction, n is the number of charge transferred, F is the Faraday constant, and M is the molecular weight of active material used in the electrode. For example,

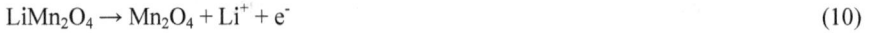

$$LiMn_2O_4 \rightarrow Mn_2O_4 + Li^+ + e^- \tag{10}$$

In this case, $\xi = 1$, $n = 1$, $F = 96500$ C mol^{-1} and $M = 180.821$ g mol^{-1}. Finally, a theoretical capacity of 148 mAh g^{-1} can be obtained for $LiMn_2O_4$.

However, the practical specific capacity of an electrode in a real LIB is usually different from the theoretical one, especially at high current, when the rate of ions diffusion across the interface between electrode and electrolyte becomes limited. A loss of the Li ions diffused into the electrode will result in a loss of capacity. Therefore, the practical capacity of the active material should be calculated as follows according to the testing results:

$$Q = \int_0^{\Delta t} I dt = \int_0^Q dq \tag{11}$$

Since the constant current mode (CC) is commonly used, in this case, the practical capacity can be obtained by $Q = It$ directly.

2.4 Coulombic Efficiency

The percent Coulombic efficiency of a LIB is defined as follows:

$$\eta_c = Q_{dis}/Q_{ch} \times 100\% \tag{12}$$

It presents the ratio between the capacity a LIB provided to the external circuit for power supply during discharge and the capacity it restored during charging. It is a very important parameter for LIBs to show their energy storage capacity. However, the Coulombic efficiency is usually not 100%. The irreversible loss of capacity is primarily due to the chemical reactions between electrode and electrolyte, electrode volume change, or electrode decomposition (dissolution). Among them, the complex electrode-electrolyte chemical reactions lead to the irreversible formation of a layer of solid electrolyte interphase (SEI) on the electrode during initial charge of a LIB, which is passivating and will reduce the reversible capacity.

2.5 Energy Density

The energy capacity is defined as the following relation:

$$W = Q \times V = \int_0^{\Delta t} IV(t)\mathrm{d}t = \int_0^Q V(q)\mathrm{d}q \qquad (13)$$

where t is time and q represents the state of charge.

Thus, the gravimetric and volumetric energy density can be calculated by

$$\text{W' (Wh kg}^{-1}) = W/m \;; \text{W' (Wh L}^{-1}) = W/V \qquad (14)$$

The relation suggests that enhancement in LIB voltage is an effective way to increase the LIB energy. Since the potential of carbonaceous anode, φ_{an} (vs. Li/Li$^+$), is close to 0, the cell voltage, E_{cell}, is usually governed by the cathode potential, φ_{cath}. In addition, choosing electrodes with high capacity also results in high energy. Thus, designing and developing high voltage and high capacity cathode materials have attracted worldwide attention.

2.6   Power Density

Similarly,

$$P(q) = W/t = I \times V \qquad (15)$$

The gravimetric and volumetric power density can be obtained as follows

$$\text{P' (W kg}^{-1}) = P/m \;; \text{P' (W L}^{-1}) = P/V \qquad (16)$$

Different from energy density which represents how long a LIB can power a load, power density shows how quickly the LIB can deliver energy. To start the engine of vehicles, power density matters.

2.7   Electrolyte Requirement LIB voltage, E_{cell}, is not governed by electrolyte. However, the use of improper electrolyte may deteriorate LIB performance since the electrolyte will experience unwanted reactions at the electrode surface during operation. In thermodynamics, the stability of electrolyte requires the potentials of the electrolyte redox reactions to be not in the LIB voltage window, as shown in Fig. 2. μ_A and μ_C are the electrochemical potentials of anode and cathode, respectively. The window of the electrolyte, E_g, is the energy gap between its lowest unoccupied (LUMO) molecular orbital and highest occupied (HOMO) one. As illustrated in Fig. 2, an anode whose electrochemical potential μ_A is higher than the LUMO will reduce the electrolyte until a passivating SEI layer forms to block electron transportation from the anode to the electrolyte LUMO while a cathode whose electrochemical potential μ_C is lower than the HOMO will oxidize the electrolyte until a passivating SEI layer builds a barrier to electron transportation from the electrolyte HOMO to the cathode. Thus, electrolyte stability requires the electrochemical potentials of anode μ_A and cathode μ_C locating within the window of the electrolyte in thermodynamics, which enforces the open-circuit voltage E_{cell} of a LIB to

$$eE_{cell} = \mu_A - \mu_C \leq E_g \qquad (17)$$

where e is the magnitude of the electron charge.

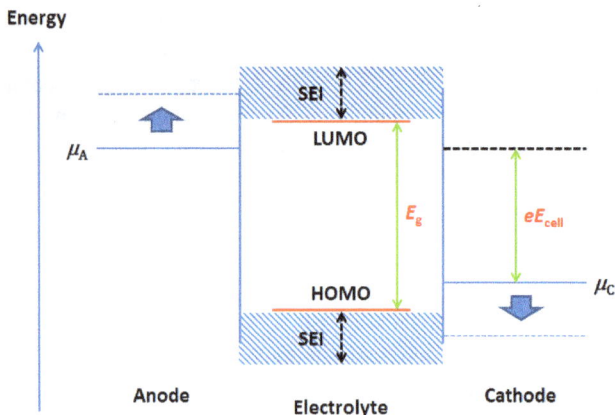

Figure 2. *Schematic energy diagram of an organic electrolyte without electrode/electrolyte reaction. E_g is the window of the electrolyte for thermodynamic stability. A μ_A > LUMO and/or a μ_C < HOMO requires a kinetic stability by the formation of an SEI layer.*

3. Recent Development of LIBs

Energy economy based on fossil fuels is currently at risk due to the depletion of non-renewable resources and the emissions of greenhouse gas CO_2. With an growing energy crisis and global temperature rise issue, the demand for clean and sustainable energy sources is becoming more and more critical. To utilize clean energy effectively, efficient EES systems are fairly essential. Among various well-developed EES systems, rechargeable LIBs are widely used in portable devices (cell phones, laptops, etc.) nowadays due to their high efficiency, high energy density and non-pollution properties.

The earliest rechargeable LIBs which used chalcogenides (TiS_2, MoS_2) as the cathodes with metallic lithium as the anodes were discovered in 1970s.[1] The structure of chalcogenides comprise various layers of hexagonal close-packed octahedral atomic groups, in which a layer of Ti, Mo atoms located in between two layers of S atoms. During discharge, Li ions insert into the structure and occupy the vacant octahedral sites between the layers. Upon charge, Li ions extract from the electrode and a reverse process occurs, maintaining the pristine chalcogenide structure. This kind of batteries, however, could only provide a low voltage of 2.2 V (V_{cell} < 2.5 V) and the metallic lithium was quickly found to be unsafe as the anode due to dendrite growth on the surface of metal after repeated Li plating, leading to internal short circuits. Since 1980 when the $LiCoO_2$ was demonstrated as a cathode material, the commercialization of the first generation LIBs by SONY, which combined $LiCoO_2$ with the carbonaceous material anodes, achieved a notable milestone. [2] This kind of batteries now dominate the portable digital market. Followed by the discovery of $LiCoO_2$, research interest for cathodes has also been put on other layered-structure $LiMO_2$ (M: Co, Mn, and/or Ni), olivine-structure $LiFePO_4$ and spinel-structure $LiMn_2O_4/LiMn_{1.5}Ni_{0.5}O_4$ materials. For anodes, TiO_2, $Li_4Ti_5O_{12}$, Sn, SnO_2, Fe_2O_3, Si and Sn-Co alloy have been developed. With the development of electrode materials, solid electrolyte and advanced high performance electrolyte showing high voltage limit have also been investigated. Nowadays LIBs is expected to support green transportations such as electric vehicles (EVs) and hybrid electric vehicles (HEVs). The increase in the demand of these vehicles requires LIBs with high energy density, high power density and good cycle performance.

4. Limitations of High Power/Energy Applications

Since Mitsubishi and General Motors first introduced EVs and HEVs powered by Li-ion batteries in early 2000s, the era of electric vehicles has dawned. During the last few decades, energy and environmental challenges have further aroused great interest in an electric transportation infrastructure, which can replace gasoline engine and reduce pollution.[3] As it is known, rechargeable batteries used in EVs include lead-acid, Ni-Cd, nickel metal hydride, lithium ion, Li-ion polymer and, less commonly, zinc-air and molten salt batteries. Among all types of batteries, LIBs are the most attractive power source for such devices due to their high energy density, high efficiency and non-pollution. They are dominating the most recent group of EVs in development as shown in Table 1.[4]

Unfortunately, the applications of EVs and HEVs have not yet found great success on a commercial scale. For example, the first generation EVs can only run for 100 km per single charge using pure electric mode, and a full charge usually takes about 8-12 hours. Moreover, compared to conventional vehicles, their acceleration performance and climbing capacity are poor.[5] The main deficiency is due to the low power and limited energy densities of the applied LIBs, as well as their insufficient life.[6, 7]

Table 1.   A comparison of different batteries for EVs and HEVs

Battery types	Energy density [Wh kg^{-1}]	Efficiency	Lifespan
Lead-acid	30-40	70-75%	3 years
Nickel metal hydride	30-80	60-70%	160,000+ km
Molten salt	120	-	1000+ cycles
Lithium ion	150+	80-90%	Hundreds to a few thousand cycles

Power density (W kg^{-1}) is defined as operation current density (A kg^{-1}) times operation voltage (V).

POWER DENSITY (W kg^{-1}) = CURRENT DENSITY (A kg^{-1}) × POTENTIAL (V)   (18)

while the amount of energy density (W h kg^{-1}) is a function of the capacity (A h kg^{-1}) and cell voltage (V).

ENERGY DENSITY (W h kg^{-1}) = CAPACITY (A h kg^{-1}) × POTENTIAL (V)   (19)

As mentioned, potential V_{cell} of a LIB is determined by the difference between φ_{cath} and φ_{an}. Since the potential of carbonaceous anode, φ_{an} (vs. Li/Li$^+$), is close to 0. As a result, V_{cell} may be enhanced only by increasing φ_{cath}. Furthermore, capacity of a LIB is currently limited by the cathodes as well. Thus, to improve the power density, both the operation current density and working voltage of the cathode materials should be increased. Meanwhile, to get higher energy density, either the capacity or working voltage of the cathode materials should be improved. Therefore, a cathode material with high working potential, good rate capability, high capacity and sufficient life is expected for high power/energy devices.

5.   Overview of Cathode Materials

Developing cathode materials with high energy density and high power density is one of the key challenges for LIBs. As discussed, high energy density and high power density can be obtained by high voltage and/or high capacity. Many kinds of cathode materials have been commercialized so far. In this section, several representative compounds will be studied, such as layered $LiCoO_2$, olivine $LiFePO_4$, spinel $LiMn_2O_4$ and their derivatives, to check their possibility for high energy/high power applications.[8] Each of them presents advantages and disadvantages as shown in Table 2.

Table 2.   Comparison of several commercial cathode materials for LIBs

Cathode material	$LiCoO_2$	$LiFePO_4$	$LiMn_2O_4$
Operating voltage [V]	3.9	3.5	4.0
Capacity [mAh g^{-1}]	130	170	148
Rate capability	2D	1D	3D
Cost	High	Low	Low
Safety	Not stable & toxicity	Most stable & non-toxicity	Non-toxicity

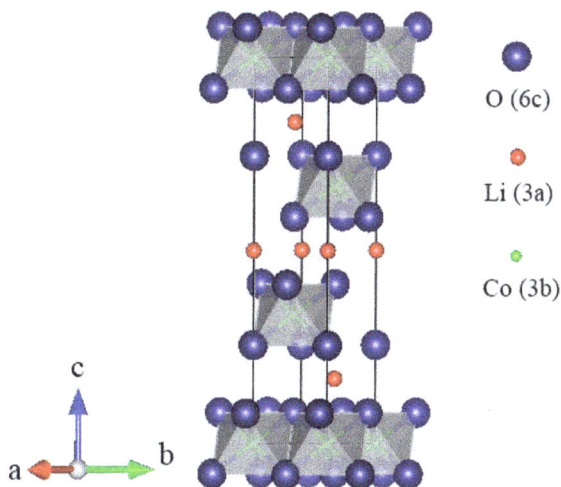

Figure 3. Crystal structure of layer $LiCoO_2$ with space group R-3m

Layered-structure $LiCoO_2$ was first discovered in 1980 by Mizushima et al.[2]. LIBs using $LiCoO_2$ now dominate more than 90% of the world's market.[9, 10] It has the α-$NaFeO_2$ layered structure with the oxygens in a cubic close-packed arrangement, similar to the structures of the dichalcogenides, in which O-M-O layers are bonded by Li ion in between them (Fig. 3).[11] This structure offers 2-dimensional (2D) Li ion diffusion pathways. Although Ohzuku et al. [12] demonstrated that layered materials might be easy to release internal stress accompanied with Li ions extraction and insertion process compared to the materials with 1D and 3D framework structures, it has been reported that only 0.5 Li can be reversibly extracted from and inserted back into $LiCoO_2$ without dramatic structure transformation, showing theoretical capacity of only 130 mAh g^{-1}. Further, Li deintercalation was found to induce a nonuniform volume change from the monoclinic phase to a hexagonal phase as well as unsafe oxygen evolution and HF attack from electrolyte due to the formation of Co^{4+}.[12-15] To improve the structural stability and enlarge the reversible capacity of LiCoO$_2$, Al and Mg were utilized to partially substitute Co in LiCo$_{1-y}$M$_y$O$_2$. However, the results were unsatisfactory.[16, 17] Besides, Co is expensive and toxic. Low reversible capacity, high cost and safety problems make it urgent to be replaced by others. Based on the template of layered LiCoO$_2$, some other layered-structure cathode materials have been studied: LiNiO$_2$, LiMnO$_2$, LiMn$_{0.5}$Ni$_{0.5}$O$_2$, LiNi$_{1/3}$Mn$_{1/3}$Co$_{1/3}$O$_2$. Among them, LiMn$_{0.5}$Ni$_{0.5}$O$_2$ and LiNi$_{1/3}$Mn$_{1/3}$Co$_{1/3}$O$_2$ showed promising electrochemical behaviors and were proposed as possible alternatives to LiCoO$_2$. [18, 19] LiMn$_{0.5}$Ni$_{0.5}$O$_2$ was able to deliver 200 mAh g^{-1} when charge and discharge between 2.5 and 4.5 V at low current density as reported by Makimura and Ohzuku.[20] However, Li/Ni mixing is always observed, which will block Li transportation pathway deteriorating the electrochemical performance of LiMn$_{0.5}$Ni$_{0.5}$O$_2$.[21, 22] Yoshio et al.[23] revealed that the addition of Co could significantly reduce the transition-metal content in the lithium layer of LiMn$_{1-y}$Ni$_y$O$_2$. Ohzuku et al.[18] first proposed LiNi$_{1/3}$Mn$_{1/3}$Co$_{1/3}$O$_2$ in 2001. They quickly found this material could exhibit a maximum capacity of about 200 mAh g^{-1} between 2.5 and 4.6 V similar to LiMn$_{0.5}$Ni$_{0.5}$O$_2$ but better rate capability, which can be associated with the reduced Li/Ni cation mixing.[24] Thus, LiNi$_{1/3}$Mn$_{1/3}$Co$_{1/3}$O$_2$ become a hot research topic in recent years. Most of these studies showed increased capacity when increasing charge potential. However, this kind of material tends to lose Li$^+$ and O^{2-} like other layered systems at potential higher than 4.6 V,[25, 26] which may lead to a large irreversible capacity. To further improve the structural stability of LiMO$_2$, Argonne National Laboratory proposed to embed structural stabilizer Li$_2$MnO$_3$ into layered materials.[27] Li$_2$MnO$_3$ has a layered rock salt structure (C2/m symmetry). As reported by Thackeray et al.[27, 28], it closely resembles the ideal layered structures of LiCoO$_2$ and can be represented in a conventional layered LiMO$_2$ notation as Li[Li$_{0.33}$Mn$_{0.67}$]O$_2$. Although Li$_2$MnO$_3$ possesses Li/Mn layers (1:2) rather than pure transition metal layers that commonly exist in LiMO$_2$, it shows a similar interlayer spacing of 4.7 Å to LiMO$_2$. The compatibility of the close-packed layers in these two compounds allows the integration of Li$_2$MnO$_3$ with LiMO$_2$ at the atomic level. As a result, a capacity of > 200 mAh g^{-1} has been achieved for xLi$_2$MnO$_3$•(1-x)LiMO$_2$ (M = Mn, Ni, Co) in high voltage 3.0-4.5 V LIBs, in which Li$_2$MnO$_3$ acts as a Li reservoir to supply extra lithium ion to adjacent lithium-depleted layer in LiMO$_2$, prevent oxygen loss during charge, and retain the structural stability. When the electrochemical potential is raised above 4.5V during charge, part of Li$_2$MnO$_3$ component was reported to be activated by further extraction of

Li ion from the Li_2MnO_3 structure with the simultaneous release of O (Li_2O). The obtained electrochemically active component MnO_2 could contribute the final capacity of the product with the formation of $LiMnO_2$. However, the coulombic efficiency is relatively low in this case since with every two Li ions extracted from Li_2MnO_3 only one Li ion can be inserted back in the following discharge process. Moreover, in most cases the problem of high cost and toxicity caused by Co still exists. Extensive researches are needed to lower the cost and toxicity of layered materials, and to optimize the high capacity materials for future applications.

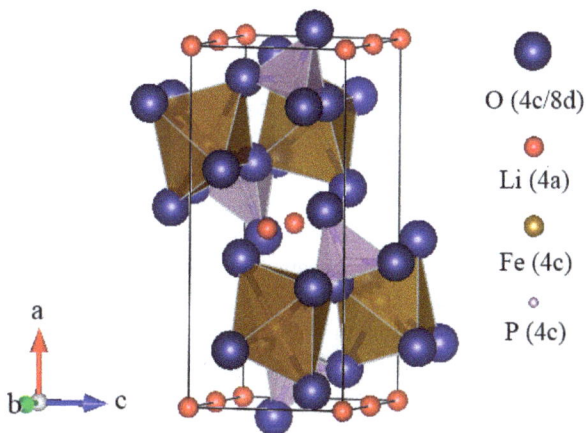

Figure 4. Crystal structure of olivine LiFePO₄ with space group Pnma

Olivine-structured $LiFePO_4$ becomes another hot commercial cathode in recent years due to its low cost, high safety and less toxicity. Padhi et al.[29] first reported this type of materials in 1997. It shows an orthorhombic lattice structure with space group *Pnma*, in which edge-shared LiO_6 octahedra and corner-shared FeO_6 octahedra packed along the *b*-axis and linked together by the PO_4 tetrahedra as shown in Fig. 4. With such structure characteristics, Li diffusion in olivine $LiFePO_4$ is believed to be 1D along the *b*-axis. Upon delithiation, one mole of Li ions are extracted to yield $FePO_4$ with the theoretical capacity of 170 mAh g^{-1} at 3.5 V. Since the oxygen atoms are bonded strongly by Fe and P atoms, the structure of $LiFePO_4$ is much more stable than layered oxides at both high temperature calcination and high potential during charging.[30-32] Such high structural stability for $LiFePO_4$ leads to good cycle performance and safety property. However, the strong bondings also result in sluggish ionic transportation (10^{-13} to 10^{-16} cm^2 s^{-1}) [33] and low electronic conductivity (10^{-9} cm s^{-1}) [34] 1D lithium pathway is another reason for the poor conductivities of $LiFePO_4$. Many researches have been conducted to improve the rate capability of $LiFePO_4$.[35, 36] Carbon coating was considered to be an effective way to enhance the electronic conductivity of olivine materials by Dominko et al.[35] and Roberts et al.[36]. On the other hand, high Li^+ mobility along the [010] or b axis has been found by extensive studies.[37-39] Decreasing particles' size[40] or only b axis[41] has been investigated and improved electrochemical performance was achieved. However, the processing cost for $LiFePO_4$ with carbon coating or small particle size is generally high. Besides, relatively low operation voltage for $LiFePO_4$ (3.5 V) limit its wide application in high energy and high power systems. In this case, other olivine structure materials $LiMPO_4$ (M = Mn, Co, Ni) have been considered, whose working potentials are increased to 4.1 V for $LiMnPO_4$, 4.8 V for $LiCoPO_4$ and 5.1 V for $LiNiPO_4$.[42-47] All these materials were born in the same work as $LiFePO_4$ by Padhi et al. in 1997.[29] $LiMnPO_4$ is the most promising candidate due to its abundant precursors, high theoretical capacity (170 mAh g^{-1}), moderately high working voltage (4.1 V) and environmental friendliness. Delacourt et al. [48] revealed that the poor ionic and electronic conductivities limited the electrochemical reaction of $LiMnPO_4$ and there was a difference of about five orders of magnitude between the conductivities of $LiFePO_4$ and $LiMnPO_4$. Similar to $LiFePO_4$, carbon coating and diminishing the particles' size have been utilized to improve the conductivities of $LiMnPO_4$. The results, however, remain controversial. The carbon-containing $LiMnPO_4$ prepared by Li et al.[49] through a solid-state reaction delivered 140 mAh g^{-1} at 0.28 mA cm^{-2}. Delacourt et al.[50] obtained ~100 nm particles of $LiMnPO_4$ by a direct precipitation route, which showed only a low capacity of 70 mAh g^{-1} at 0.05 C. Kwon et al.[51] synthesized $LiMnPO_4$/C with particle size of about 130 nm using a sol-gel method followed by ballmilling, which delivered 134 mAh g^{-1} at 0.1 C. All the results suggest that it is difficult to approach the theoretical capacity of $LiMnPO_4$. Extensive researches are still needed for this material to obtain optimized electrochemical performance. Besides, $LiCoPO_4$ and $LiNiPO_4$ also suffer from poor conductivities leading to low capacity. Moreover, their working voltage, especially that of $LiNiPO_4$, is too high for current electrolyte. Therefore, enhancing the rate capability of $LiMnPO_4$ is an important task for application of olivine structure materials in advanced vehicles.

Spinel material $LiMn_2O_4$ has attracted much attention because of its advantages of abundant precursor (12th most abundant element on earth), low cost (five times cheaper than Co), non-toxicity, high operation voltage (4.0 V) and good rate performance (3D framework). Thackeray et al. [52] first proposed $LiMn_2O_4$ in 1984. It was reported to have a face-centered cubic spinel structure $A[B]_2X_4$ with space group of Fd-$3m$ (Fig. 5), where cation Li occupies the tetragonal site and cation Mn occupies half of the octahedral site. The left half empty octahedral site surrounds the tetragonal site where Li stays and provides a 3D channels for Li diffusion during Li ions extraction and insertion reactions. [53-56] $LiMn_2O_4$ is usually studied between 3.0 and 4.0 V. In this case, lithium ions can be extracted from $LiMn_2O_4$ to form the λ-MnO_2 phase, resulting in a theoretical reversible capacity of approximately 148 mAh g^{-1} and an energy density (> 500 Wh kg^{-1}) comparable to that of $LiCoO_2$.[61, 62] Lithium-ion insertion into λ-MnO_2 causes the cell to expand isotropically up to the composition $LiMn_2O_4$, maintaining cubic structure. Despite their advantages of precursor abundance, easy preparation, low cost, and non-toxicity, the application of spinel materials is still limited. Three major disadvantages plague spinel cathodes for industrial application.

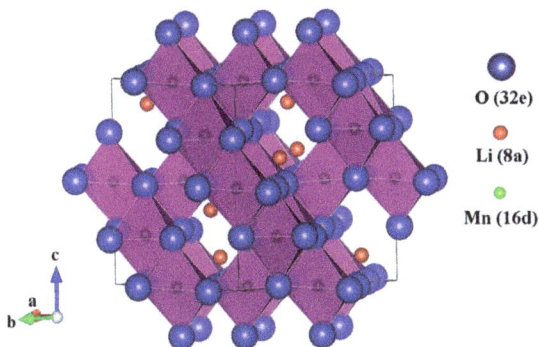

O (32e)

Li (8a)

Mn (16d)

Figure 5. Crystal structure of LiMn₂O₄ spinel with space group Fd-3m

(i) poor rate capability

Pristine spinel materials have been reported to have low electronic and ionic conductivities, as well as slow diffusion of lithium ions at the electrode/electrolyte interface.[82] For example, $LiMn_2O_4$ has low electrical conductivity of 10^{-6} S cm^{-1} and lithium-ion diffusion coefficient of ~ 10^{-11} cm^2 s^{-1} accordingly.[76, 83] The low electric conductivity of $LiMn_2O_4$ can limit the current flow between particles while the low lithium-ion diffusion coefficient can limit the Li ions extraction/insertion at the electrode/electrolyte interface.

(ii) Mn^{3+} ions dissolution

In spite of poor rate capability, fast capacity loss of $LiMn_2O_4$ cells is another major reason for its limited application. Many evidences point out that solubility of the spinel electrode in the electrolyte at the top of charge is one of the reasons for the capacity fade in spinel cells.[84-89] Accordingly, the solubility of charged electrodes has been attributed to a disproportionation reaction at the particle surface:[84]

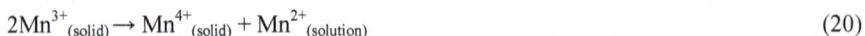

$$2Mn^{3+}_{(solid)} \rightarrow Mn^{4+}_{(solid)} + Mn^{2+}_{(solution)} \tag{20}$$

The formation of the Mn^{3+} ions at the top of a charge is mainly associated with oxygen loss from the spinel lattice caused by the reaction with electrolyte.[90]

On the other hand, since the dominant lithium salt for lithium-ion batteries is $LiPF_6$, which is sensitive to a trace amount of moisture, $LiPF_6$ will react with traces of water to produce acids (HF) according to the following reaction

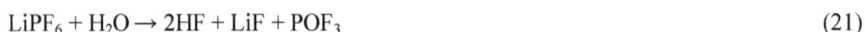

$$LiPF_6 + H_2O \rightarrow 2HF + LiF + POF_3 \tag{21}$$

As can be seen, with the presence of moisture, $LiPF_6$ is readily decomposed to form an HF-containing acidic environment, which significantly accelerates the dissolution of manganese ions.[91-92]

(iii) Jahn-Teller distortion

Jahn-Teller distortion which occurs in spinel material during cycling is another important reason for its capacity fade.

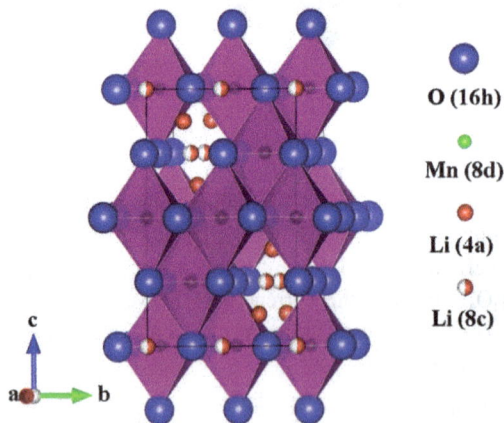

Figure 6. The tetragonal $Li_2Mn_2O_4$ structure (space group: $I4_1/amd$)

Jahn-Teller effect is a phase transition process for spinel system from cubic crystal structure to tetragonal $Li_2Mn_2O_4$ with additional lithium ions insertion into $LiMn_2O_4$. Fig.

6 shows the detail atomic configuration in a unit cell of $Li_2Mn_2O_4$ which has a tetragonal structure with a space group of $I4_1/amd$. In detail, the 4a, 8c, 8d and 16h positions in the tetragonal structure are corresponding to the 8a, 16c, 16d and 32e positions in the cubic spinel structure with $a_T \approx a_C/\sqrt{2}$ and $c_T \approx a_C$. As can be seen, there is 16% anisotropy in the lattice parameters on formation of the tetragonal phase.[54, 93-94] The anisotropic volume change along with the cubic-tetragonal phase transition results in a loss of capacity on repeated cycling.[89] Although Jahn-Teller distortion usually occurs when the spinel electrode is discharged below 3 V, the formation of $Li_2Mn_2O_4$ is also found at the spinel electrode surface at the end of a discharge of 3 V, particular under high rate.[84, 89, 95-96] The existence of $Li_2Mn_2O_4$ at the surface is due to kinetic limitations during fast intercalation. Under dynamic, some crystallites can be more lithiated than others, thereby driving the composition of the electrode surface into a Mn^{3+}-rich region. High concentration of Jahn-Teller active Mn^{3+} changes the structure from cubic to tetragonal with accompanied structural stress, leading to capacity fade. Moreover, the Mn^{3+}-rich regions are subject to dissolution and the resulting phase boundary separation between cubic and tetragonal regions can all contribute to a rapid capacity fading.[97] As a result, the effect of Jahn-Teller distortion should be taken into consideration and eliminated no matter in what potential window the spinel $LiMn_2O_4$ cell works.

To solve the problems caused by Mn^{3+} in $LiMn_2O_4$, high voltage spinels $LiMn_{2-x}M_xO_4$ with Mn in $LiMn_2O_4$ substituted by other transition metals, such as Ni^{2+}[63-66], Co^{3+}[67-69] and Fe^{3+}[70-73], were studied for high energy systems. Among them, $LiMn_{1.5}Ni_{0.5}O_4$ is the most successful and attractive candidate, whose working voltage is as high as 4.7 V but still barely within the limit of conventional organic electrolyte. Besides, good cycle stability could be achieved in this case since the oxidation valence of Mn is +4 after substituting Mn^{3+} with Ni^{2+} when cycled above 3 V. Oh et al. [4] reported that the origin of the high voltage plateau in the $LiMn_{1.5}Ni_{0.5}O_4$ oxide is attributed to the difference (~0.5 eV) in binding energy of the top valence band between nickel Ni^{2+} 3d $e_g(\uparrow\downarrow)$ and the manganese Mn^{3+} 3d $e_g(\uparrow)$ level. The impressively high voltage makes its energy density (650 Wh kg^{-1}) 20% and 30% higher than that of conventional $LiCoO_2$ and $LiFePO_4$ materials, respectively, making it potential to be deployed in HEVs and PHEVs.[74] In terms of $LiMn_{1.5}Ni_{0.5}O_4$, it has two different crystal structures with different arrangements of Mn^{4+} and Ni^{2+} ions in the lattice.[78-79] The atoms arrangement of $LiMn_{1.5}Ni_{0.5}O_4$ in the face-centered cubic spinel structure with Fd-$3m$ space group is very like pure spinel $LiMn_2O_4$, which is also known as disordered spinel. In this case, the Ni and Mn atoms are randomly distributed in the 16d sites as shown in Fig. 5. In contrast, in the primitive simple cubic spinel structure with $P4_332$ space group (ordered spinel), the Ni, Mn, and Li atoms are respectively occupied in the 4b, 12d, and

8c sites and O ions are occupied in the 8c and 24e sites with highly cation ordering (Fig. 7).

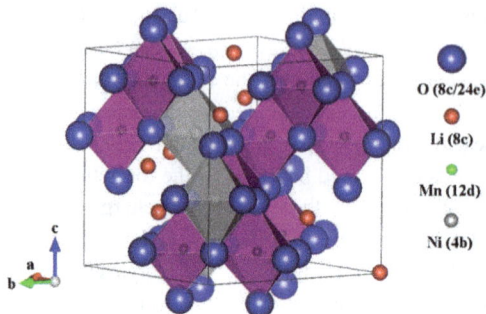

Figure 7. Crystal structure of LiMn$_{1.5}$Ni$_{0.5}$O$_4$ spinel with space group P4$_3$32

Although the two structures both have the 3D lithium diffusion channels, they are a little different from each other. For LiMn$_2$O$_4$ and disordered LiMn$_{1.5}$Ni$_{0.5}$O$_4$ (*Fd-3m*), Li diffuses by moving from an 8a site to the neighboring empty octahedral 16c site, and then to the next 8a site, while in ordered LiMn$_{1.5}$Ni$_{0.5}$O$_4$ (*P4$_3$32*), the octahedral vacant 16c sites are split into ordered 4a and 12d sites and thus, Li diffuses by partially moving to 4a and the rest to 12d sites. Xia et al.[80] found the disorder spinel LiMn$_{1.5}$Ni$_{0.5}$O$_4$ displays a high Li diffusivity throughout the full composition range and is comparable to that of layered LiCoO$_2$. Meanwhile, Liu and his co-workers[81] pointed out that ordering of the Ni and Mn retards the lithium diffusivity and lowers the rate capability. In summary, the disorder spinel oxide is expected to exhibit better rate capability than order spinel oxide. Moreover, poor rate capability, surface side reaction and cubic to tetragonal transition are also faced by LiMn$_{1.5}$Ni$_{0.5}$O$_4$ and the Jahn-Teller distortion seems more complicated. For example, Wagemaker et al.[98] investigated the electrochemical behavior in the 3 V region through the combination of the X-ray and neutron diffraction analysis and found the extensive migration of Ni and Mn during cycling, resulting in the disappearance of the initial Ni-Mn ordering and the formation of Ni-rich and Ni-poor domains, resulting in two Jahn-Teller distorted tetragonal phases with different Ni:Mn ratios (T1 and T2 phases). Anyway, Lee et al.[7] revealed that a large volume change accompanied with cubic to tetragonal transitions both in ordered and disordered LiMn$_{1.5}$Ni$_{0.5}$O$_4$ systems.

Improvement in the rate capability and cyclability of LiMn$_2$O$_4$/LiMn$_{1.5}$Ni$_{0.5}$O$_4$ spinel is important for their application in high power/energy vehicles. Most strategies, such as nanosizing, surface coating and doping, have been utilized to solve these problems and previous studies are summarized in Table 3. Considerable results have been achieved in rate performance enhancement and material dissolution reduction for spinel systems. However, the detail mechanisms behind the Jahn-Teller distortion still remain controversial and its improvement is thus unsatisfactory. More researches should be carried out.

Table 3.   *Comparison of several commercial cathode materials for LIBs*

Target	Strategy	Current research state
Better rate capability	Nanosizing	Nanobelts, nanotubes, nanorods, nanowires, nanoparticles
	Doping	Zn^{2+}, $Cu^{2+/3+}$, Co^{3+}, Mg^{2+}, Nd^{3+}, Y^{3+}
Better capacity retention	Surface coating	$Li_2O \bullet 2B_2O_3$ glass, MgO, Al_2O_3, CeO_2, SnO_2, SiO_2, $LiCoO_2$, ZrO_2/Li_2ZrO_3, ZnO, [Li, La]TiO_3, $Li_4Ti_5O_{12}$, $ZnMn_2O_4$, $FePO_4$, acetylene black, gold, silver, polypyrrole
	Doping	Li^+, Ni^{2+}, Cr^{3+}, Co^{3+}, Ti^{4+}, Al^{3+}, B^{3+}, Nd^{3+}, Zn^{2+}, $Cu^{2+/3+}$, Fe^{3+}, Mg^{2+}, Ca^{2+}, Ga^{3+}, Y^{3+}, Sr^{2+}, cation-anion codoping (Al-S, Cr-F, Co-F, Al-F)

6. Conclusions

Rechargeable batteries are a good choice for electrical energy storage ranging from portable electronics to transportation vehicles, such as electric vehicle and hybrid electric vehicle. In order to meet the requirements of high energy and high power applications, recent development of lithium-ion batteries have been reviewed. Researches on layered-, olivine- and spinel- structured cathode materials for lithium-ion batteries are now turning into new standard.

References

[1]   M.S. Whittingham, Electrical Energy Storage and Intercalation Chemistry, Science 192 (1976) 1126-1127. https://doi.org/10.1126/science.192.4244.1126

[2]   K. Mizushima, P.C. Jones, P.J. Wiseman, J.B. Goodenough, $Li_xCoO_2(0<x<1)$: a new cathode material for batteries of high energy density, Mat. Res. Bull. 15 (1980) 783-789. https://doi.org/10.1016/0025-5408(80)90012-4

[3]   R. Santhanam, B. Rambabu, Research progress in high voltage spinel $LiNi_{0.5}Mn_{1.5}O_4$ material, J. Power Sources 195 (2010) 5442-5451. https://doi.org/10.1016/j.jpowsour.2010.03.067

[4]   S.H. Oh, S.H. Jeon, W.I. Cho, C.S. Kim, B.W. Cho, Synthesis and characterization of the metal-doped high-voltage spinel $LiNi_{0.5}Mn_{1.5}O_4$ by mechanochemical process, J. Alloys Compd. 452 (2008) 389-396. https://doi.org/10.1016/j.jallcom.2006.10.153

[5]   H. Wang, 4d Transition metal doped $LiNi_{0.5}Mn_{1.5}O_4$ cathodes for high power lithium batteries, National University of Singapore, 2011.

[6]   Z. Chen, Y. Qin, K. Amine, Y.K. Sun, Role of surface coating on cathode materials for lithium-ion batteries, J. Mater. Chem. 20 (2010) 7606-7612. https://doi.org/10.1039/c0jm00154f

[7]   E.-S. Lee, K.-W. Nam, E. Hu, A. Manthiram, Influence of Cation Ordering and Lattice Distortion on the Charge–Discharge Behavior of $LiMn_{1.5}Ni_{0.5}O_4$ Spinel between 5.0 and 2.0 V, Chem. Mater. 24 (2012) 3610-3620. https://doi.org/10.1021/cm3020836

[8]   X. Ma, B. Kang, G. Ceder, High Rate Micron-Sized Ordered $LiNi_{0.5}Mn_{1.5}O_4$, J. Electrochem. Soc. 157 (2010) A925-A931. https://doi.org/10.1149/1.3439678

[9]   M.S. Whittingham, Lithium batteries and cathode materials, Chemical Reviews 104 (2004) 4271-4301. https://doi.org/10.1021/cr020731c

[10] T. Ohzuku, R.J. Brodd, An overview of positive-electrode materials for advanced lithium-ion batteries, J. Power Sources 174 (2007) 449-456. https://doi.org/10.1016/j.jpowsour.2007.06.154

[11] A. Mendiboure, C. Delmas, P. Hagenmuller, New layered structure obtained by electrochemical deintercalation of the metastable $LiCoO_2$ (O2) variety, Mat. Res. Bull. 19 (1984) 1383-1392. https://doi.org/10.1016/0025-5408(84)90204-6

[12] T. Ohzuku, A. Ueda, Why transition metal (di)oxides are the most attractive mateirals for batteries, Solid State Ionics 69 (1994) 201-211. https://doi.org/10.1016/0167-2738(94)90410-3

[13] T.A. Arunkumar, E. Alvarez, A. Manthiram, Structural, Chemical, and Electrochemical Characterization of Layered $Li[Li_{0.17}Mn_{0.33}Co_{0.5-y}Ni_y]O_2$ Cathodes, J. Electrochem. Soc. 154 (2007) A770-A775. https://doi.org/10.1149/1.2745635

[14] J.B. Goodenough, Y. Kim, Challenges for Rechargeable Li Batteries, Chem. Mater. 22 (2010) 587-603. https://doi.org/10.1021/cm901452z

[15] J.N. Reimer, J.R. Dahn, Electrochemical and in situ X-ray diffraction studies of lithium intercalation in Li_xCoO_2, J. Electrochem. Soc. 8 (1992) 2091-2097. https://doi.org/10.1149/1.2221184

[16] Y.-I. Jang, B. Huang, H. Wang, D.R. Sadoway, G. Ceder, Y.M. Chiang, H. LIu, H. Tamura, $LiAl_yCo_{1-y}O_2$ (R-3m) intercalation cathode for rechargeable lithium batteries, J. Electrochem. Soc. 146 (1999) 862-868. https://doi.org/10.1149/1.1391693

[17] H. Tukamoto, A.R. West, Electronic conductivity of $LiCoO_2$ and its enhancement by magnesium doping, J. Electrochem. Soc. 144 (1997) 3164-3168. https://doi.org/10.1149/1.1837976

[18] T. Ohzuku, Y. Makimura, Layered Lithium Insertion Material of $LiCo_{1/3}Ni_{1/3}Mn_{1/3}O_2$ for Lithium-Ion Batteries, Chemistry Letters 30 (2001) 642-643. https://doi.org/10.1246/cl.2001.642

[19] T. Ohzuku, Y. Makimura, Layered Lithium Insertion Material of $LiNi_{1/2}Mn_{1/2}O_2$: A Possible Alternative to $LiCoO_2$ for Advanced Lithium-Ion Batteries, Chemistry Letters 30 (2001) 744-745. https://doi.org/10.1246/cl.2001.744

[20] Y. Makimura, T. Ohzuku, Lithium insertion material of $LiNi_{1/2}Mn_{1/2}O_2$ for advanced lithium-ion batteries, J. Power Sources 119-121 (2003) 156-160. https://doi.org/10.1016/S0378-7753(03)00170-8

[21]   K. Kang, Y.S. Meng, J. Breger, C.P. Grey, G. Ceder, Electrodes with high power and high capacity for rechargeable lithium batteries, Science 311 (2006) 977-980. https://doi.org/10.1126/science.1122152

[22]   S.B. Schougaard, J. Bréger, M. Jiang, C.P. Grey, J.B. Goodenough, $LiNi_{0.5+\delta}Mn_{0.5-\delta}O_2$—A High-Rate, High-Capacity Cathode for Lithium Rechargeable Batteries, Adv. Mater. 18 (2006) 905-909. https://doi.org/10.1002/adma.200500113

[23]   M. Yoshio, H. Noguchi, J.-i. Itoh, M. Okada, T. Mouri, Preparation and properties of $LiCo_yMn_xNi_{1-x-y}O_2$ as a cathode for lithium ion batteries, J. Power Sources 90 (2000) 176-181. https://doi.org/10.1016/S0378-7753(00)00407-9

[24]   N. Yabuuchi, T. Ohzuku, Novel lithium insertion material of $LiCo_{1/3}Ni_{1/3}Mn_{1/3}O_2$ for advanced lithium-ion batteries, J. Power Sources 119-121 (2003) 171-174. https://doi.org/10.1016/S0378-7753(03)00173-3

[25]   Z. Lu, L.Y. Beaulieu, R.A. Donaberger, C.L. Thomas, J.R. Dahn, Synthesis, Structure, and Electrochemical Behavior of $Li[Ni_xLi_{1/3-2x/3}Mn_{2/3-x/3}]O_2$, J. Electrochem. Soc. 149 (2002) A778-A791. https://doi.org/10.1149/1.1471541

[26]   Z. Lu, J.R. Dahn, Understanding the Anomalous Capacity of $Li/Li[Ni_xLi_{(1/3-2x/3)}Mn_{(2/3-x/3)}]O_2$ Cells Using In Situ X-Ray Diffraction and Electrochemical Studies, J. Electrochem. Soc. 149 (2002) A815-A822. https://doi.org/10.1149/1.1480014

[27]   M.M. Thackeray, S.-H. Kang, C.S. Johnson, J.T. Vaughey, R. Benedek, S.A. Hackney, Li_2MnO_3-stabilized $LiMO_2$ (M = Mn, Ni, Co) electrodes for lithium-ion batteries, J. Mater. Chem. 17 (2007) 3112-3125. https://doi.org/10.1039/b702425h

[28]   M.M. Thackeray, C.S. Johnson, J.T. Vaughey, H.N. LiCurrent address: eVionyx Inc, S.A. Hackney, Advances in manganese-oxide 'composite' electrodes for lithium-ion batteries, J. Mater. Chem. 15 (2005) 2257. https://doi.org/10.1039/b417616m

[29]   A.K. Padhi, K.S. Nanjundaswamy, J.B. Goodenough, Phospho-olivines as positive-electrode materials for rechargeable lithium batteries, J. Electrochem. Soc. 144 (1997) 1188-1194. https://doi.org/10.1149/1.1837571

[30]   G. Arnold, J. Garche, R. Hemmer, S. Ströbele, C. Vogler, M. Wohlfahrt-Mehrens, Fine-particle lithium iron phosphate $LiFePO_4$ synthesized by a new low-cost aqueous precipitation technique, J. Power Sources 119-121 (2003) 247-251. https://doi.org/10.1016/S0378-7753(03)00241-6

[31] M. Takahashi, S.-i. Tobishima, K. Takei, Y. Sakurai, Reaction behavior of LiFePO$_4$ as a cathode material for rechargeable lithium batteries, Solid State Ionics 148 (2002) 283-289. https://doi.org/10.1016/S0167-2738(02)00064-4

[32] J.R. Dahn, E.W. Fuller, Thermal stability of Li$_x$CoO$_2$, Li$_x$NiO$_2$ and λ-MnO$_2$ and consequences for the safety of Li-ion cells, Solid State Ionics 69 (1994) 265-270. https://doi.org/10.1016/0167-2738(94)90415-4

[33] X.-C. Tang, L.-X. Li, Q.-L. Lai, X.-W. Song, L.-H. Jiang, Investigation on diffusion behavior of Li$^+$ in LiFePO$_4$ by capacity intermittent titration technique (CITT), Electrochim. Acta 54 (2009) 2329-2334. https://doi.org/10.1016/j.electacta.2008.10.065

[34] C. Wang, J. Hong, Ionic/Electronic Conducting Characteristics of LiFePO$_4$ Cathode Materials, Electrochem. Solid-State Lett. 10 (2007) A65-A69. https://doi.org/10.1149/1.2409768

[35] R. Dominko, M. Bele, M. Gaberscek, M. Remskar, D. Hanzel, S. Pejovnik, J. Jamnik, Impact of the Carbon Coating Thickness on the Electrochemical Performance of LiFePO$_4$/C Composites, J. Electrochem. Soc. 152 (2005) A607-A610. https://doi.org/10.1149/1.1860492

[36] M.R. Roberts, G. Vitins, J.R. Owen, High-throughput studies of Li$_{1-x}$Mg$_{x/2}$FePO$_4$ and LiFe$_{1-y}$Mg$_y$PO$_4$ and the effect of carbon coating, J. Power Sources 179 (2008) 754-762. https://doi.org/10.1016/j.jpowsour.2008.01.034

[37] M.S. Islam, D.J. Driscoll, C.A.J. Fisher, P.R. Slater, Atomic-scale investigation of defects, dopants, and lithium transport in the LiFePO$_4$ olivine-type battery material, Chem. Mater. 17 (2005) 5085-5092. https://doi.org/10.1021/cm050999v

[38] T. Maxisch, F. Zhou, G. Ceder, Ab initiostudy of the migration of small polarons in olivine Li$_x$FePO$_4$ and their association with lithium ions and vacancies, Phys. Rev. B 73 (2006) 104301. https://doi.org/10.1103/PhysRevB.73.104301

[39] R. Amin, P. Balaya, J. Maier, Anisotropy of Electronic and Ionic Transport in LiFePO$_4$ Single Crystals, Electrochem. Solid-State Lett. 10 (2007) A13-A16. https://doi.org/10.1149/1.2388240

[40] C. Delacourt, P. Poizot, S. Levasseur, C. Masquelier, Size Effects on Carbon-Free LiFePO$_4$ Powders, Electrochem. Solid-State Lett. 9 (2006) A352-A355. https://doi.org/10.1149/1.2201987

[41] K. Dokko, S. Koizumi, H. Nakano, K. Kanamura, Particle morphology, crystal orientation, and electrochemical reactivity of LiFePO$_4$ synthesized by the

hydrothermal method at 443 K, J. Mater. Chem. 17 (2007) 4803-4810. https://doi.org/10.1039/b711521k

[42]   D. Morgan, A. Van der Ven, G. Ceder, Li Conductivity in Li_xMPO_4 (M=Mn, Fe, Co, Ni) Olivine Materials, Electrochem. Solid-State Lett. 7 (2004) A30-A32. https://doi.org/10.1149/1.1633511

[43]   H. Fang, Z. Pan, L. Li, Y. Yang, G. Yan, G. Li, S. Wei, The possibility of manganese disorder in $LiMnPO_4$ and its effect on the electrochemical activity, Electrochem. Commun. 10 (2008) 1071-1073. https://doi.org/10.1016/j.elecom.2008.05.010

[44]   M. Yonemura, A. Yamada, Y. Takei, N. Sonoyama, R. Kanno, Comparative Kinetic Study of Olivine Li_xMPO_4 (M=Fe, Mn), J. Electrochem. Soc. 151 (2004) A1352-A1356. https://doi.org/10.1149/1.1773731

[45]   K. Amine, H. Yasuda, M. Yamachi, Olivine $LiCoPO_4$ as 4.8 V electrode material for lithium batteries, Electrochem. Solid-State Lett. 3 (2000) 178-179. https://doi.org/10.1149/1.1390994

[46]   M.E. Rabanal, M.C. Gutierrez, F. Garcia-Alvarado, E.C. Gonzalo, M.E. Arroyo-de Dompablo, Improved electrode characteristics of olivine–$LiCoPO_4$ processed by high energy milling, J. Power Sources 160 (2006) 523-528. https://doi.org/10.1016/j.jpowsour.2005.12.071

[47]   C.M. Julien, A. Mauger, K. Zaghib, R. Veillette, H. Groult, Structural and electronic properties of the $LiNiPO_4$ orthophosphate, Ionics 18 (2012) 625-633. https://doi.org/10.1007/s11581-012-0671-6

[48]   C. Delacourt, L. Laffont, R. Bouchet, C. Wurm, J.B. Leriche, M. Morcrette, J.M. Tarascon, C. Masquelier, Toward Understanding of Electrical Limitations (Electronic, Ionic) in $LiMPO_4$ (M=Fe, Mn) Electrode Materials, J. Electrochem. Soc. 152 (2005) A913-A921. https://doi.org/10.1149/1.1884787

[49]   G. Li, H. Azuma, M. Tohda, $LiMnPO_4$ as the Cathode for Lithium Batteries, Electrochem. Solid-State Lett. 5 (2002) A135-A137. https://doi.org/10.1149/1.1475195

[50]   C. Delacourt, P. Poizot, M. Morcrette, J.M. Tarascon, C. Masquelier, One-step low-temperature route for the preparation of electrochemically active $LiMnPO_4$ powders, Chem. Mater. 16 (2004) 93-99. https://doi.org/10.1021/cm030347b

[51]   N.-H. Kwon, T. Drezen, I. Exnar, I. Teerlinck, M. Isono, M. Graetzel, Enhanced Electrochemical Performance of Mesoparticulate $LiMnPO_4$ for Lithium Ion

Batteries, Electrochem. Solid-State Lett. 9 (2006) A277-A280.
https://doi.org/10.1149/1.2191432

[52] M.M. Thackeray, P.J. Johnson, L.A.d. Picciotto, Electrochemical extractin of
      lithium from LiMn$_2$O$_4$, Mater. Res. Bull. 19 (1984) 179-187.
      https://doi.org/10.1016/0025-5408(84)90088-6

[53] K.-H. Hwang, S.-H. Lee, S.-K. Joo, Characterization of sputter-deposited
      LiMn2O4 thin films for rechargeable microbatteries, J. Electrochem. Soc. 141
      (1994) 3296-3299. https://doi.org/10.1149/1.2059329

[54] T. Ohzuku, M. Kitagawa, T. Hirai, Electrochemistry of manganese dioxide in
      lithium nonaqueous cell, J. Electrochem. Soc. 137 (1990) 769-775.
      https://doi.org/10.1149/1.2086552

[55] I. Yamada, T. Abe, Y. Iriyama, Z. Ogumi, Lithium-ion transfer at LiMn$_2$O$_4$ thin
      film electrode prepared by pulsed laser deposition, Electrochem. Commun. 5
      (2003) 502-505. https://doi.org/10.1016/S1388-2481(03)00113-9

[56] S. Kobayashi, I.R.M. Kottegoda, Y. Uchimoto, M. Wakihara, XANES and
      EXAFS analysis of nano-size manganese dioxide as a cathode material for
      lithium-ion batteries, J. Mater. Chem. 14 (2004) 1843-1848.
      https://doi.org/10.1039/b315443b

[57] X. Li, Y. Xu, Enhanced cycling performance of spinel LiMn$_2$O$_4$ coated with
      ZnMn$_2$O$_4$ shell, J. Solid State Electrochem. 12 (2007) 851-855.
      https://doi.org/10.1007/s10008-007-0426-x

[58] X. Li, Y. Xu, C. Wang, Suppression of Jahn–Teller distortion of spinel LiMn2O4
      cathode, J. Alloys Compd. 479 (2009) 310-313.
      https://doi.org/10.1016/j.jallcom.2008.12.081

[59] A. Van der Ven, C. Marianetti, D. Morgan, G. Ceder, Phase transformations and
      volume changes in spinel LixMn2O4, Solid State Ionics 135 (2000) 21-32.
      https://doi.org/10.1016/S0167-2738(00)00326-X

[60] G. Amatucci, A.D. Pasquier, A. Blyr, T. Zheng, J.M. Tarascon, The elevated
      temperature performance of the LiMn$_2$O$_4$/C system: failure and solutions,
      Electrochemica Acta 45 (1999) 255-271. https://doi.org/10.1016/S0013-
      4686(99)00209-1

[61] K.M. Shaju, P.G. Bruce, A stoichiometric nano-LiMn2O4 spinel electrode
      exhibiting high power and stable cycling, Chem. Mater. 20 (2008) 5557-5562.
      https://doi.org/10.1021/cm8010925

[62]   S.H. Park, K.S. Park, Y.-K. Sun, K.S. Nahm, Synthesis and characterization of a new spinel, Li1.02Al0.25Mn1.75O3.97S0.03, operating at potentials between 4.3 and 2.4 V, J. Electrochem. Soc. 147 (2000) 2116-2121. https://doi.org/10.1149/1.1393494

[63]   Y.-K. Sun, D.-W. Kim, Y.-M. Choi, Synthesis and characterization of spinel $LiMn_{2-x}Ni_xO_4$ for lithium_polymer battery applications, J. Power Sources 79 (1999) 231-237. https://doi.org/10.1016/S0378-7753(99)00160-3

[64]   Y. Xia, T. Sakai, T. Fujieda, M. Wada, H. Yoshinaga, A 4 V Lithium-Ion Battery Based on a 5 V $LiNi_xMn_{2-x}O_4$ Cathode and a Flake Cu-Sn Microcomposite Anode, Electrochem. Solid-State Lett. 4 (2001) A9-A11. https://doi.org/10.1149/1.1339238

[65]   Y.J. Wei, L.Y. Yan, C.Z. Wang, X.G. Xu, F. Wu, G. Chen, Effects of Ni doping on $[MnO_6]$ octahedron in $LiMn_2O_4$, J. Phys. Chem. B 108 (2004) 18547-18551. https://doi.org/10.1021/jp0479522

[66]   Y. Wei, K. Nam, K. Kim, G. Chen, Spectroscopic studies of the structural properties of Ni substituted spinel LiMn2O4, Solid State Ionics 177 (2006) 29-35. https://doi.org/10.1016/j.ssi.2005.10.015

[67]   P. Arora, B.N. Popov, R.E. White, Electrochemical investigations of cobalt-doped LiMn2O4 as cathode material for lithium-ion batteries, J. Electrochem. Soc. 145 (1998) 807-815. https://doi.org/10.1149/1.1838349

[68]   H. Kawai, M. Nagata, H. Kageyama, H. Tukamoto, A.R. West, 5 V lithium cathodes based on spinel solid solutions $Li_2Co_{1+x}Mn_{3-x}O_8$, Electrochim. Acta 45 (1999) 315-327. https://doi.org/10.1016/S0013-4686(99)00213-3

[69]   C.H. Shen, R.S. Liu, R. Gundakaram, J.M. Chen, S.M. Huang, J.S. Chen, C.M. Wang, Effect of Co doping in $LiMn_2O_4$, J. Power Sources 102 (2001) 21-28. https://doi.org/10.1016/S0378-7753(01)00765-0

[70]   H. Kawai, M. Nagata, M. Tabuchi, H. Tukamoto, A.R. West, Novel 5 V spinel cathode $Li_2FeMn_3O_8$ for lithium ion batteries, Chem. Mater. 10 (1998) 3266-3268. https://doi.org/10.1021/cm9807182

[71]   T. Ohzuku, K. Ariyoshi, S. Takeda, Y. Sakai, Synthesis and characterization of 5 V insertion material of $Li[Fe_yMn_{2-y}]O4$ for lithium-ion batteries, Electrochim. Acta 46 (2001) 2327-2336. https://doi.org/10.1016/S0013-4686(00)00725-8

[72]   A. Eftekhari, Electrochemical performance and cyclability of $LiFe_{0.5}Mn_{1.5}O_4$ as a 5 V cathode material for lithium batteries, J. Power Sources 124 (2003) 182-190. https://doi.org/10.1016/S0378-7753(03)00602-5

[73]   S.M. Malyovanyi, A.A. Andriiko, A.P. Monko, Synthesis and electrochemical behavior of Fe-doped overstoichiometric $LiMn_2O_4$ -based spinels, J. Solid State Electrochem. 8 (2003) 7-10. https://doi.org/10.1007/s10008-003-0414-8

[74]   J. Xiao, X. Chen, P.V. Sushko, M.L. Sushko, L. Kovarik, J. Feng, Z. Deng, J. Zheng, G.L. Graff, Z. Nie, D. Choi, J. Liu, J.G. Zhang, M.S. Whittingham, High-performance $LiNi_{0.5}Mn_{1.5}O_4$ spinel controlled by Mn^{3+} concentration and site disorder, Adv. Mater. 24 (2012) 2109-2116. https://doi.org/10.1002/adma.201104767

CHAPTER 2

High Potential LiNi$_{0.5}$Mn$_{1.5}$O$_4$ Cathode for LIBs

Hailong Wang

School of Physics and Electronic-Electrical Engineering, Ningxia University

merrick_whl@126.com

Abstract

Spinel-structured LiNi$_{0.5}$Mn$_{1.5}$O$_4$ is a unique cathode material for lithium ion batteries (LIBs), which has drawn great attentions owing to its high working potential (> 4.5 V vs. Li$^+$/Li) as well as large reversible capacity (> 130mAh.g^{-1}) . However, rate capability and cyclic performance of the LiNi$_{0.5}$Mn$_{1.5}$O$_4$ still need significant improvement before it can be truly used in real applications. Intensive research works have shown that its crystal structure, conductivity, concentration of Mn^{3+} ions, surface chemistry and particle morphology can significantly contribute to its electrochemical performances.

Keywords

LiNi$_{0.5}$Mn$_{1.5}$O$_4$, Crystal Structure, Conductivity, Surface Coating, Particle Morphology

Contents

1.   Introduction

Increasing demands for lithium ion batteries with higher power density and longer work-life have stimulated researchers to develop cathode materials with larger reversible capacity, higher working potential and a more stable crystal structure [1,2]. Decades of research have developed several types of cathode materials with different advantages compared with the traditional $LiCoO_2$ cathode, such as layered structured $LiNi_{1/3}Co_{1/3}Mn_{1/3}O_2$ with expanded capacity, spinel structured $LiMn_2O_4$ with higher working potential and Olivine structured $LiFePO_4$ with quite stable crystal structure. $LiNi_{0.5}Mn_{1.5}O_4$ is a derivative of $LiMn_2O_4$, which maintains the basic spinel crystal structure by substituting Mn^{3+} ions with Ni^{2+} ions. It shows a promising working potential around 4.7 V vs Li^+/Li, which is much higher than that of $LiCoO_2$ (3.8 V), $LiMn_2O_4$ (4.1 V) and $LiFePO_4$ (3.4 V) but still within the electrochemical window of $LiPF_6$ based electrolyte [2]. Such high potential could render the battery to have a much higher power output at the same current density; meanwhile its crystal structure is more stable than that of the $LiMn_2O_4$ as detrimental Mn^{3+} ions are almost absent, and its reversible capacity is greater than 130 $mAh.g^{-1}$ [3]. All above advantages make the $LiNi_{0.5}Mn_{1.5}O_4$ a promising candidate for high power applications. However, electrochemical tests show that its rate performance and cyclic performance are not satisfactory yet for real applications. This chapter provides a review of extensive works carried out on basic properties and improvement studies of the spinel structured $LiNi_{0.5}Mn_{1.5}O_4$.

2.   Basic Properties and Crystal Structure

$LiNi_{0.5}Mn_{1.5}O_4$ spinels can have either one of the two different structures with space group belonging to $Fd\bar{3}m$ and $P4_332$, respectively. Both of them possess the cubic structure as the basic framework, and the major difference between them lies in the way of metal ions' distribution [4,5]. As illustrated in Fig. 1, for the $Fd\bar{3}m$ phase, Li ions occupy the tetrahedral 8a sites, Ni and Mn ions randomly distribute in the octahedral 16d sites; while in the $P4_332$ phase, Li ions stay in the 8c sites, Ni ions occupy the 4a sites and Mn ions stay in the 12d sites [4,6]. High resolution X-ray diffractions (XRD) scan is a powerful way to discern the $LiNi_{0.5}Mn_{1.5}O_4$ spinels with different space groups. Comparing the XRD profile of the $LiNi_{0.5}Mn_{1.5}O_4$ with the $Fd\bar{3}m$ space group, extra weak peaks at around 15.3°, 39.7°, 45.7°, 57.5°and 65.6° are present in the XRD profile of the sample with $P4_332$ space group due to high degree of cation ordering at octahedral sites[7-9]. Fourier transformed infrared (FTIR) and Raman spectra have also been demonstrated as effective methods to discern the two different phases [10-12].

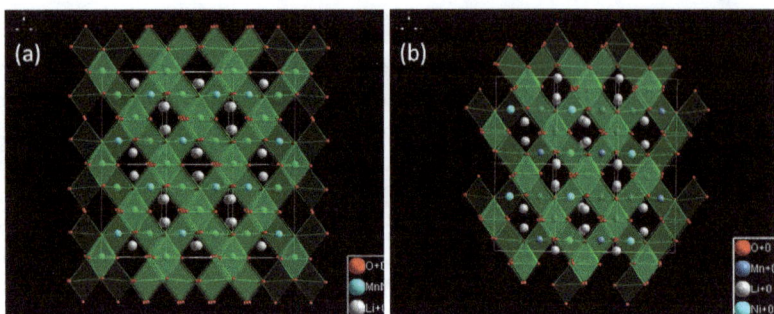

Figure 1. Crystal structure of LiNi0.5Mn1.5O4: (a) Fd$\bar{3}$m and (b) $P4_332$.

Many research works have systematically evaluated and compared the electrochemical performances of the $LiNi_{0.5}Mn_{1.5}O_4$ with different space groups. Generally, at low current densities (<0.2 C, 1 C= 147 mA.g^{-1}) of both two types of $LiNi_{0.5}Mn_{1.5}O_4$ show similar reversible capacity of about 135 mAh.g^{-1} with major voltage plateau at 4.7 V. Fd$\bar{3}$m $LiNi_{0.5}Mn_{1.5}O_4$ exhibits a short plateau at around 4.0 V contributing by $Mn^{3+/4+}$ redox, while, all Mn ions in the $P4_332$ $LiNi_{0.5}Mn_{1.5}O_4$ are tetravalent [13-16]. However, by further increasing the discharge rate to 1-6 C significant difference in accessible capacity can be observed. The discharge capacity of $P4_332$ $LiNi_{0.5}Mn_{1.5}O_4$ drops to 80 mAh.g^{-1} or less at a high rate of about 3 C, and its voltage plateau also drops to the region between 4.0 and 4.5 V. In contrast, Fd$\bar{3}$m $LiNi_{0.5}Mn_{1.5}O_4$ can still offer about 120 mAh.g^{-1} at the same condition with major voltage plateau falling within the region between 4.25 and 4.6 V [4,17,18]. The two types of $LiNi_{0.5}Mn_{1.5}O_4$ show similar cycling performances at low rates < 0.2 C, where both of them can deliver about 130 mAh.g^{-1} with slow capacity fading. When cycling at about 3 C, $P4_332$ $LiNi_{0.5}Mn_{1.5}O_4$ delivers less than 100 mAh.g^{-1} at the initial few cycles and the capacity fading is very fast, while Fd$\bar{3}$m $LiNi_{0.5}Mn_{1.5}O_4$ initially can still offer a large capacity of nearly 120 mAh.g^{-1} with slow capacity fading. Further increase the cycling rate to around 10 C, the performance of Fd$\bar{3}$m $LiNi_{0.5}Mn_{1.5}O_4$ worsens too with little initial capacity of about 60 mAh.g^{-1} and quickly drops to less than 40 mAh.g^{-1} [3,19,20]. The excellent rate performance and stable cycling performance of cathode materials are of utmost importance for high power LIBs. Neither Fd$\bar{3}$m $LiNi_{0.5}Mn_{1.5}O_4$ nor the one with $P4_332$ space group exhibits a satisfactory high rate performance. Investigating their intrinsic properties could help understand the poor high rate performances of $LiNi_{0.5}Mn_{1.5}O_4$ spinels, and also could help pinpoint directions for improvement work. The crystal structure evolution during Li^+ ions extraction/insertion process, and the structure related conductivity of $LiNi_{0.5}Mn_{1.5}O_4$ spinels are investigated. Ex-situ XRD study of the structure evolution of $LiNi_{0.5}Mn_{1.5}O_4$ shows that the lattice constant of $Fd\bar{3}m$ $LiNi_{0.5}Mn_{1.5}O_4$ shrinks as extracting Li^+ ions [21]. A new cubic phase appears when 50% of Li^+ ions are extracted, and this phase gradually becomes the dominant phase at the fully charged state. While structure evolution is more complex for the $P4_332$ $LiNi_{0.5}Mn_{1.5}O_4$, three different cubic phases appear at different Li content ranges with the last one dominating at the fully charged state. In-situ XRD and neutron diffraction data have been also used to analyze the structural evolution of $LiNi_{0.5}Mn_{1.5}O_4$ during charge/discharge process [22-25], and similar phase transition phenomenon have been confirmed. It is generally believed that the complicated phase-transition gives rise to poor rate capabilities. Further studies have revealed that poorer conductivities of $P4_332$ $LiNi_{0.5}Mn_{1.5}O_4$ could be the root cause of its worse rate capabilities compared to that of $Fd\bar{3}m$ $LiNi_{0.5}Mn_{1.5}O_4$. Kim et al. have compared area-specific impedances (ASIs) of the two types of $LiNi_{0.5}Mn_{1.5}O_4$ at the different state-of-charge (SOC), and proposed that smaller ASIs of $Fd\bar{3}m$ $LiNi_{0.5}Mn_{1.5}O_4$ result in better high rate performances compared with $P4_332$ $LiNi_{0.5}Mn_{1.5}O_4$ [21]. However, Kunduraci's research has pointed out that the disparity of the surface conductivity tends to diminish after several charge/discharge cycles, and therefore it would not lead to significant difference in electrochemical performances of the two $LiNi_{0.5}Mn_{1.5}O_4$ spinels; instead, about 1.5 orders of magnitude difference in the intrinsic electronic conductivity has been found through *dc* polarization measurements, and the better high rate performances of $Fd\bar{3}m$ $LiNi_{0.5}Mn_{1.5}O_4$ have been attributed to better intrinsic electronic conductivity [17,18,22]. Wang's research on high rate performances of $LiNi_{0.5}Mn_{1.5}O_4$ further revealed that poor intrinsic conductivity could result in accumulation of detrimental Mn^{3+} J-T ions , which are present to balance the piling up of positively charged Li^+ ions at particle's surface; high concentration of Mn^{3+} ions gradually deteriorate the particles' integrity and significantly worsen the cycling performance [19,26]. In order to improve the high rate performances of $LiNi_{0.5}Mn_{1.5}O_4$ spinels, many methods have been exploited, including reduction of particles' size, cation/anion doping and surface modifications.

3.   Electrochemical Performances

As stated in above sections, $LiNi_{0.5}Mn_{1.5}O_4$ exhibits different electrochemical performances due to differences in crystal structure, Mn^{3+} ion concentration, intrinsic conductivity, surface chemistry and particle morphology. Among these factors, crystal structure, Mn^{3+} ions concentration, intrinsic conductivity and surface chemistry have been found to have strong correlation. In addition, influence of particle morphology on high rate performances has been also found to have strong impact on the performance of batteries.

3.1 Effect of Mn^{3+} Ions

Mn^{3+} ions have been identified as a key factor that could influence electrochemical performances of many cathode materials. Mn^{3+} ion has strong tendency to dissolve into electrolyte through a disproportionate reaction: $2Mn^{3+} \rightarrow Mn^{4+} + Mn^{2+}$, where the Mn^{2+} dissolves into electrolyte and leave Mn^{4+} on the particle's surface, which results in corrosion of particles and deterioration of conductivity [27,28]. Moreover, Mn^{3+} is a well-known Jahn-Teller ion, which can cause significant lattice distortion during electrochemical cycling, and eventually leads to particle breakdown and loss of electrical contact with current collector [1,29]. Suffered from above negative effects, the famous spinel structured $LiMn_2O_4$ cathode shows quite poor cyclic performance since half of its Mn ions are trivalent [30,31].

By using Ni^{2+} ions to substitute Mn^{3+} ions, the stoichiometric and phase pure $LiNi_{0.5}Mn_{1.5}O_4$ can be free of Mn^{3+} ions, and thus better electrochemical performances compared with the $LiMn_2O_4$ can be expected. However, many researches have proven that the crystal structure of $LiNi_{0.5}Mn_{1.5}O_4$ is very sensitive to its thermal treatment history, which in turn alters its electrochemical performance [27,32-40]. Usually well-crystallized $LiNi_{0.5}Mn_{1.5}O_4$ is obtained by calcination at above 650 °C. If the final calcination temperature is below 800 °C (around 700 °C), the obtained product would possess the stoichiometric element ratio, perfect spinel structure with $P4_332$ space group which is free of Mn^{3+} ions. While, if the final calcination temperature is above 800 °C, the as-synthesized product will possess a defected spinel structure, in which few oxygen ions escaped from the lattice during the high temperature process and a trace amount of Li_xNiO_x impurity can be detected by high resolution XRD scan. To balance the loss of negative charged oxygen ions, a small amount of Mn^{4+} ions transform into Mn^{3+} ions, which together with Mn^{4+} and Ni^{2+} ions are randomly distributed in the octahedral sites (16d) manifesting a space group of $Fd\bar{3}m$.

With free of Mn^{3+} ions, the crystal structure of $LiNi_{0.5}Mn_{1.5}O_4$ ($P4_332$ phase) would be more stable compared with the $Fd\bar{3}m$ phase. At low charge/discharge rates (≤ 0.1 C) it indeed shows large reversible capacity and excellent cyclic performance with a small capacity fading. While the $LiNi_{0.5}Mn_{1.5}O_4$ ($Fd\bar{3}m$ phase) shows faster capacity fading at the same current density which can be ascribed to the small amount of detrimental Mn^{3+} ions. At increased rates at about 1 C, the electrochemical performances of the

$LiNi_{0.5}Mn_{1.5}O_4$ ($P4_332$ phase) deteriorate significantly with smaller reversible capacity and worse cyclic performance compared with the $LiNi_{0.5}Mn_{1.5}O_4$ ($Fd\bar{3}m$ phase) because Mn^{3+} could be a double edged sword, detrimental for crystal stability, but beneficial for conductivity [27,41,42]. Many research works have shown that the $LiNi_{0.5}Mn_{1.5}O_4$ ($Fd\bar{3}m$ phase) with trace amount of Mn^{3+} ions exhibits smaller polarization during charge/discharge compared with the one with a $P4_332$ space group suggesting a better conductivity of the $LiNi_{0.5}Mn_{1.5}O_4$ ($Fd\bar{3}m$ phase). Amatucci et al. have studied their intrinsic electronic conductivities by measuring the impedance on very dense pellets, which shows that the electronic conductivity of the $LiNi_{0.5}Mn_{1.5}O_4$ ($Fd\bar{3}m$ phase) is about 10 times higher than that of the $LiNi_{0.5}Mn_{1.5}O_4$ ($P4_332$ phase), and the improved electronic conductivity is attributed to electron hoping from increased content of Mn^{3+} to Mn^{4+} [17,18,22,43]. Moreover, existence of Mn^{3+} can improve Li^+ diffusion from two aspects: firstly, the ionic radius of Mn^{3+} is larger than that of the Mn^{4+}, which gives rise to a larger lattice volume to lower the activation energy of Li^+ hopping [19]; secondly, in-situ XRD study shows that Li^+ insertion/extraction in the ordered $P4_332$ phase occurs in two two-phase reactions with fixed lattice constant within certain Li^+ content range, which constrains the Li^+ diffusion significantly, while the Li^+ insertion/extraction in the disordered $Fd\bar{3}m$ phase occurs in one two-phase reaction with higher diffusivity, where the disordered cation distribution is caused by the presence of Mn^{3+} [16,25,28,44,45]. It can be seen that trace amount of Mn^{3+} is beneficial for both electrons and Li^+ ions transportation inside the crystal lattice, which explains better performances of the $LiNi_{0.5}Mn_{1.5}O_4$ ($Fd\bar{3}m$ phase) at increased rates. However, at high rates (≥ 10 C), both of the ordered and disordered $LiNi_{0.5}Mn_{1.5}O_4$ show severe polarization, drastically reduced reversible capacity and very fast capacity fading. A failure mechanism at high rates has been proposed in Fig 2 [19,26].

As shown in Fig 2, during high-rate discharge Li^+ ions are easily piled-up on surfaces of $LiNi_{0.5}Mn_{1.5}O_4$ particles due to limited conductivities of the ordered and disordered phase, leading to formation of a Li-rich phase $Li_{2-\delta}Ni_{0.5}Mn_{1.5}O_4$ ($\delta<1$) on the particle's surface. To balance the increased positive charge from excess Li^+ ions on the surfaces, some Mn^{4+} ions transform into Mn^{3+} to keep charge neutrality. The detrimental J-T distortion becomes more pronounced with increased Mn^{3+} concentration. Upon cycling, particles suffering from large stress variation will gradually crack and break down, and lose contact with conductive carbon and current collector which aggravates the polarization. In addition, Mn^{3+} tends to react with electrolyte and dissolve into it, therefore high concentration of Mn^{3+} on the surface will make the dissolution effect more severe, and further deteriorate the high rate performances.

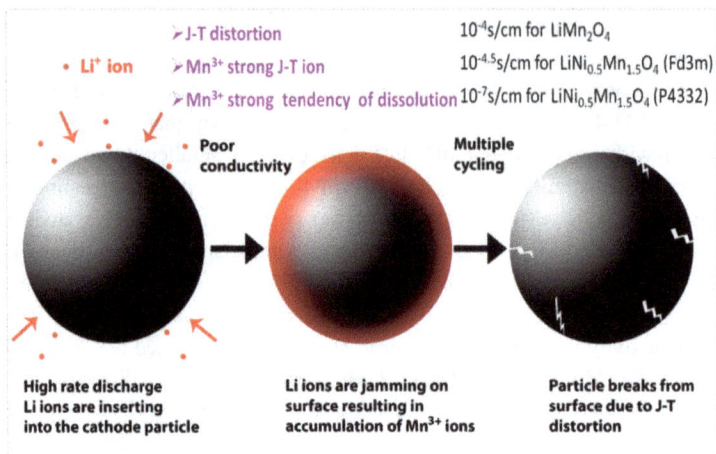

Figure 2. Illustration of failure mechanism of LiNi$_{0.5}$Mn$_{1.5}$O$_4$ particle at high rates.

3.2   Effect of Doping

Cation doping is an effective method to modify intrinsic properties of LiNi$_{0.5}$Mn$_{1.5}$O$_4$. Partial substitution of Ni or Mn by other cations such as Mg, Al, Zn, Cu, Cr, Co, Fe, V, Sm, Ga and Ru can improve the electrochemical performances through different mechanisms [11,12,19,46-64]. Anion doping is the other way to improve the performances especially the cyclic performances [65-68].

Careful crystal structure studies of Mg doped LiNi$_{0.5}$Mn$_{1.5}$O$_4$ by neutron diffraction show that LiMg$_\delta$Ni$_{0.5-\delta}$Mn$_{1.5}$O$_4$ ($0 \leqslant \delta \leqslant 0.10$) spinels possess a cation ordered structure, and interestingly, extensive migration of Mn and Ni ions occurs in these spinels during Li$^+$ insertion process which in turn forms two Ni-rich and Ni-poor domains exhibiting different electrochemical performances [11,69]. Substitutions of Al for both Ni and Mn in LiNi$_{0.5}$Mn$_{1.5}$O$_4$ have demonstrated the ability of enhancing cyclic performance and thermal stability, which could be attributed to a stronger Al-O bond compared with Ni-O and Mn-O bond [53]. Zn doped LiNi$_{0.42}$Mn$_{1.5}$Zn$_{0.08}$O$_4$ exhibits improved cyclic performance with >97% capacity retention in 50 cycles compared to 92% for LiNi$_{0.5}$Mn$_{1.5}$O$_4$. The smaller lattice parameter difference of three cubic phases formed during cycling has been proposed as the origin of the improvement through Zn doping [70]. Cu doping could give rise to a higher discharge plateau at 4.95 V as well as higher discharge capacities at high rates, which are results of new Cu$^{2+/3+}$ redox and higher lithium diffusion coefficient after Cu doping, respectively [51,52]. A small amount of Cr doping has been reported to be beneficial for rate capabilities and elevated-temperature performances. Strong oxygen affinity of Cr, absence of second cubic phase during cycling and suppressing SEI layer formation after Cr doping are proposed as the main mechanisms behind these enhancements [50,54,56,57,64,71]. Co substitution for both Ni and Mn has exhibited distinguished improvements in both rate capability and cyclic performances of $LiNi_{0.5}Mn_{1.5}O_4$, attributed to smaller lattice constant variation during cycling and better kinetic properties compared to these of the pristine $LiNi_{0.5}Mn_{1.5}O_4$ [72,73]. Fe substitution for both Ni and Mn are of great interest to researchers due to low cost of raw materials. Excellent rate capabilities and cyclic performances have been found for the Fe doped $LiNi_{0.5}Mn_{1.5}O_4$. The improved electrochemical performances are considered as a result of suppression of SEI layer formation, stabilization of cation-disordered structure and improvement in electronic conductivity [12,64,70]. Substituting Li with trace amount of V ions has been interestingly found that the degree of cation ordering is increased while the conductivity is improved at the same time [61]. The partial substitution of Sm^{3+} for Mn^{3+} in $LiNi_{0.5}Mn_{1.5}O_4$ leads to a decrease in unit cell volumes which stabilizes the crystal structure, while the rate capabilities and cyclic performances have been found improved with increasing in Sm content, which can be attributed to better structural stability and better conductivity after Sm doping [62]. 4d transition-metal Ru has also been used to replace Ni ions, and outstanding rate capabilities and cyclic performances at 10 C rate have been achieved from $LiNi_{0.5-2x}Ru_xMn_{1.5}O_4$ ($0 < x \leq 0.05$); both electronic conductivity and lithium ion diffusivity of $LiNi_{0.5-2x}Ru_xMn_{1.5}O_4$ have been found to be several times higher than those of pristine $LiNi_{0.5}Mn_{1.5}O_4$, which are attributed to delocalized Ru 4d orbitals, formation of octahedral vacancies and larger lattice parameter[19]. Recently, Lee has investigated the influence of cation doping on cation ordering and lattice microstrain of $LiNi_{0.5-x}Mn_{1.5}M_xO_4$ (M=Cr, Fe, Co and Ga), suggesting that low degree of cation ordering and low microstrain could favor better electrochemical performances [64].

Fluorine doped sample are usually synthesized by annealing the mixture of metal oxides precursors and LiF or NH_4F, and trace amount of O^{2-} is replaced by F^- [65-67]. After introducing F^- ions, the degree of cation ordering is lowered and structure stability is enhanced since Mn-F and Ni-F bonds are stronger than original Mn-O and Ni-O bonds. Therefore thermal stability and cyclic performances are significantly enhanced [65-68].

Generally speaking, transitional metal doping can significantly influence the intrinsic conductivity through changes in the electronic structure and adjusting the Li^+ ions migration condition, and improved rate performances. Anion doping could usually enhance the stability of $LiNi_{0.5}Mn_{1.5}O_4$ and thus improve the cyclic performances.

3.3 Effect of Surface Coating

Although great improvements in electrochemical performances have been achieved through doping methods, cathode materials are still facing other issues when operating in conventional electrolyte, such as corrosion by HF, dissolution of metal ions from the surface, SEI layer formation and increase in charge transfer resistance [3,4,26,37,74]. These adverse effects are more pronounced for $LiNi_{0.5}Mn_{1.5}O_4$ since its main voltage plateau is close to the upper limit of conventional electrolytes' stability window, which in turn leads to worse electrochemical performances. Surface coating is an effective way to manipulate surface properties of cathode particles, and influences of Al_2O_3, Bi_2O_3, SiO_2, ZnO, ZrO_2, $FePO_4$, $AlPO_4$, $Li_4P_2O_7$, Borate, Au, Ag, carbon, polymide and graphene oxide layers on $LiNi_{0.5}Mn_{1.5}O_4$ have been studied [26,75-94].

Al_2O_3 coating layer has demonstrated capability of suppressing SEI formation and diminishing charge transfer resistance of the pristine and doped $LiNi_{0.5}Mn_{1.5}O_4$, giving rise to better cyclic performances [72,95-97]. Bi_2O_3 coating layer has the effects of suppressing electrolyte decomposition and hindering the dissolution of transition metal ions for Ti doped $LiNi_{0.5}Mn_{1.5}O_4$, and thus capacity fading can be effectively suppressed [77]. Porous SiO_2 coating layer with nanostructure could effectively protect the $LiNi_{0.5}Mn_{1.5}O_4$ particles from HF attack, and thus the improved capacity retention has been achieved at an elevated temperature of 55°C [78]. Similar to the SiO_2 coated sample, the ZnO coated $LiNi_{0.5}Mn_{1.5}O_4$ has exhibited increased discharge capacity with good capacity retention, however, the stability of ZnO coating layer during cycling is questionable [83,92]. ZrO_2 coated $LiNi_{0.5}Mn_{1.5}O_4$ also exhibit better cyclic performance compared with the bared sample, which is due to smaller surface reactivity and improvement in charge transfer resistance; but other research points out that ZrO_2 coating layers could impair the accessible capacity of $LiNi_{0.5}Mn_{1.5}O_4$ because of its poor lithium ion transportation kinetics [79,82]. Both $FePO_4$ and $AlPO_4$ coated $LiNi_{0.5}Mn_{1.5}O_4$ have exhibited remarkable cyclic performance and electrochemical reversibility, which can be attributed to the suppressed SEI layer and the enhanced surface charge conductivity [85,86]. $Li_4P_2O_7$ coated $LiNi_{0.5}Mn_{1.5}O_4$ with little Li_3PO_4 impurity has been prepared by solid state reaction, showing significantly improved cyclic performances. The $Li_4P_2O_7$ layer is able to prevent the formation of the SEI layer and increase the surface ionic conductivity [80]. $LiPO_3$ coated $LiNi_{0.5}Mn_{1.5}O_4$ exhibits reduced impedance and enhanced surface passivation properties [89]. Simply mixing Tris(trimethylsilyl) borate with particles could effectively hinder the oxidative reactions between electrolyte with $LiNi_{0.5}Mn_{1.5}O_4$ [94]. Ceramic glass LATP (Li_2O-Al_2O_3-TiO_2-P_2O_5) coating layer have been found to be a good Li^+ ions conductor as well as surface protector for the $LiNi_{0.5}Mn_{1.5}O_4$ [88]. Noble metals such as Au and Ag were also investigated as coating materials. Ag coating layer has been obtained by mixing Ag(CH$_3$-COO) and LiNi$_{0.5}$Mn$_{1.5}$O$_4$ in Tollens solution in a mole ratio of 5: 95 [98] and Au coating layer is obtained by mixing LiNi$_{0.5}$Mn$_{1.5}$O$_4$ and HAuCl$_4$ solution in a mole ratio of 1: 99 [87]. Both Au and Ag coatings are able to improve the accessible capacity by hindering electrolyte decomposition. However, both Au and Ag coating layers have shown adverse effects on high rate performances due to raised lithium diffusion energy barrier on the electrolyte/electrode interface [87,98].

Carbon coating is an effective method to enhance conductivity and alter the surface chemistry of many electrode materials [91,99,100]. Normally, carbon coating process is conducted at low temperature (400-650 °C) in air to form residue carbon layer through pyrolysis of carbon source on oxide particles' surfaces. Since the fabrication of many transition-metal-oxide based cathodes needs calcination at high temperature (700-1000 °C) in air to form well-crystallized particles, the whole coating process generally undergoes multiple steps at different temperatures. In-situ carbon coating on oxides particles at high temperature (> 700 °C) in air would cause carbon to burn out [91,101,102]. Xerogel carbon, sucrose, phenolic resin and polyethylene glycol-4000 have been used as carbon sources. These are further decomposed in air at low temperature between 300-600 °C to form a carbon coating layer on well-crystallized LiNi$_{0.5}$Mn$_{1.5}$O$_4$ particles to improve surface conductivity and passivation properties [90,103-105]. Generally, carbon coating at low temperature would alter the surface chemistry without affecting the crystal structure and intrinsic properties of the as-prepared LiNi$_{0.5}$Mn$_{1.5}$O$_4$.

A carbon coating layer can be directly obtained at high temperature over 700 °C by pyrolysis of the carbon sources under a reducing or inert atmosphere, which has been successfully used to synthesize many silicate and phosphate based composites such as Li$_2$MnSiO$_4$/C and LiFePO$_4$/C, where a reducing or inert atmosphere is compatible with these cathode materials' synthesis condition [106,107]. For most transition-metal-oxide based cathodes like LiNi$_{0.5}$Mn$_{1.5}$O$_4$, however, strong reducing atmosphere causes valence change of some transition-metal ion such as reduction of Mn^{4+} to Mn^{3+}, deteriorating electrochemical performances [91]. Thus, direct carbon coating on well-crystallized transition-metal-oxide based cathodes was generally thought as impossible.

Carbon coated LiNi$_{0.5}$Mn$_{1.5}$O$_4$ can also be achieved through one-step high temperature (800°C) calcination process, in which carbon coating is obtained in an oxygen shorted suffocating environment using polyethylene glycol-400 as carbon source. Interestingly, not only the surface conductivity is improved after the high temperature carbon coating process, but also the crystal structure is affected where the degree of cation ordering is boosted and the concentration of Mn^{3+} is lowered as shown in Fig. 3. Raman signal of amorphous carbon is detected, and extra super lattice points arising from high degree of cation ordering appears in the SAED pattern ([110] zone) of the carbon coated sample [26]. Therefore, it is proposed that very thin residue coating layer formed at high temperature hinders the loss of oxygen ions from the lattice and therefore gives rise to a high degree of cation ordering and low concentration of detrimental Mn^{3+} ions [26].

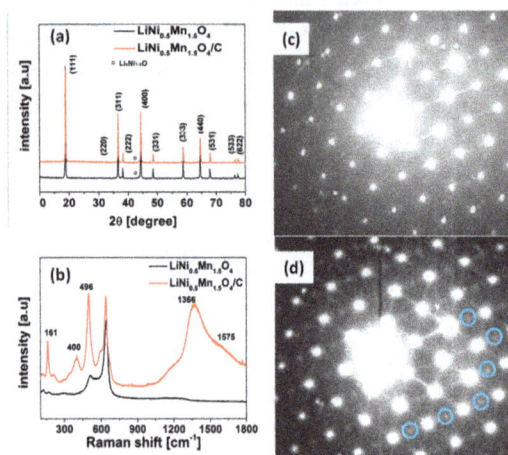

Figure 3. (a) XRD profiles of the $LiNi_{0.5}Mn_{1.5}O_4$ and the $LiNi_{0.5}Mn_{1.5}O_4/C$; (b) Raman spectra of the $LiNi_{0.5}Mn_{1.5}O_4$ and the $LiNi_{0.5}Mn_{1.5}O_4/C$; (c) SAED pattern of the $LiNi_{0.5}Mn_{1.5}O_4$ sample in [110] zone; (d) SAED pattern of the $LiNi_{0.5}Mn_{1.5}O_4/C$ sample in [110] zone **[26]**.

Nanostructure-tuned polymide coating layer has shown features of highly continuous coverage and fast ion transportation property, which could effectively alleviate undesired side reactions between electrolyte and charged $LiNi_{0.5}Mn_{1.5}O_4$ without impairing surface charge conductivity compared with metal oxide coating layers [61,108]. Recently, graphene oxide has been successfully coated on $LiNi_{0.5}Mn_{1.5}O_4$ by simply mixing and annealing method, the coating layer not only exhibits the feature of continuous coverage but also shows smaller interface charge transfer resistance, which leads to a remarkable capacity retention of 61% for 1000 cycles [109]. Graphene wrapped $LiNi_{0.5}Mn_{1.5}O_4$ with high degree of cation ordering exhibits outstanding rate capabilities, which could be attributed to the good conductive properties of graphene layers [110].

In summary, surface coating on $LiNi_{0.5}Mn_{1.5}O_4$ particles using oxides, phosphates, metals, carbon and polymers can offer benefits of suppressing the undesired SEI formation, hindering metal ions dissolution and preventing HF attack, which generally favor cyclic performance. Rate capability is very sensitive to the kinetic properties of the coating layer, and thus porous and highly conductive coating materials have exhibited promising effects of improving high rate performances. Despite of all these benefits, the weight ratio of coating materials in the electrode should be controlled within a reasonable range to avoid jeopardizing overall energy density of the electrode.

3.4   Effect of Particle Morphology

Electrode materials with nanostructure have attracted great attentions in recent years. Merits of the nano particles' size include the very short transportation length for charge carriers and the expanded area of electrode/electrolyte interfaces, which could generally enable better rate capabilities [111-115]. The characteristic time constant t of the Li^+ diffusion in the cathode particles is reduced with the square of particle size according to Eq. (1):

$$t = L^2/D \tag{1}$$

where L is the average diffusion length and D is the diffusion constant of the material, so as for the electron transportation; Meanwhile, a larger electrolyte/electrode contact area would allow a higher Li^+ flux across the interface [116].

Various methods including composite carbonate process, emulsion drying method, pechini method, hydrothermal method, polymer assisted method and so on, have been utilized to fabricate nanostructured $LiNi_{0.5}Mn_{1.5}O_4$ particles with size ranging from 20 to 300 nm [18,20,117-132]. Extensive research has shown that the calcination temperature (450 to 800 °C) is crucial for crystallinity and particle sizes. Generally speaking, a lower calcination temperature results in smaller particle sizes which favor the fast charge transportation but result in poor crystallinity, which usually could cause low accessible capacity [20,38,133]. Therefore, precise control of calcination temperature in each synthesis method is significant to balance different performance requirements.

Sucrose-aided combustion method operated at 500 °C yields $LiNi_yMn_{2-y}O_4$ $(0 < y \leq 0.5)$ with particle sizes uniformly distributed at about 22 nm, and capacity of 132.7 mAh.g^{-1} at the 2 C rate [134]. Using a polymer assisted method, well-crystallized $LiNi_{0.5}Mn_{1.5}O_4$ ($Fd\bar{3}m$) with particle size of 70 nm can be obtained at 800 °C, where the polymers act as an inhibitor for particle growth at high temperature. The as-synthesized sample delivers 98 mAh.g^{-1} at 15 C [20]. Wang's research on $Fd\bar{3}m$ type $LiNi_{0.5}Mn_{1.5}O_4$ has shown that 300 nm sized particles synthesized through PEG-400 assisted method offers 83 mAh.g^{-1} at 10 C rate. However, 1 μm sized particles can only release 58 mAh.g^{-1} at the same rate [19]. Recently, a morphology-inheritance route is developed to successfully synthesize one dimensional nanoporous $LiNi_{0.5}Mn_{1.5}O_4$ with $P4_332$ space group, and an excellent discharge capacity of 109 mAh.g^{-1} at 20 C is achieved [135]. Hydrothermal synthesized $LiNi_{0.5}Mn_{1.5}O_4$ with core-shell structure delivers a capacity of 118 mAh.g^{-1} at 10C discharge rate [136]. In-situ templated method is another way that obtains nanoflake-stacked $LiNi_{0.5}Mn_{1.5}O_4$ with an outstanding rate capability of 96 mAh.g^{-1} at an extremely high rate of 50 C [137]. The great improvement has been attributed to faster Li$^+$ diffusion between nanoflakes.

Remarkable high rate performances have been reported on the nanostructured $LiNi_{0.5}Mn_{1.5}O_4$, but it should be noted that different content and types of conductive carbon are used in above research works, which could also put significant influence on the high rate performances. In order to rule out the impact of conductive carbon, which may lead to different extrinsic electronic conductivity, Kunduraci has carefully normalized the extrinsic electronic conductivity, and compared rate capability of different particle sizes of $LiNi_{0.5}Mn_{1.5}O_4$. The comparison suggests that coarse particles (1 μm), fine particles (70 nm) could deliver more 14% and 36% in capacity capacities at 16 C rate for $Fd\bar{3}m$ and $P4_332$ space groups, respectively [18].

Although, excellent rate capabilities have been achieved through synthesis of nanostructured $LiNi_{0.5}Mn_{1.5}O_4$ particles, it is necessary to point out that smaller particle size can gives rise to adverse effects including undesired electrode/electrolyte side reactions and lower volume metric density [116]. Kunduraci has compared performances of $LiNi_{0.5}Mn_{1.5}O_4$ synthesized by pechini method at different calcination temperatures of 500, 600, 700 and 800°C respectively, and concluded that smaller particles obtained at lower temperature lead to capacity degradation and poor cyclic performance due to large electrolyte/electrode interface area associated strong side reactions [17,18,22]. Those adverse effects have to be carefully tailored or solved before putting such nanostructured $LiNi_{0.5}Mn_{1.5}O_4$ into commercial applications.

4.   Conclusion

Spinel-structured $LiNi_{0.5}Mn_{1.5}O_4$ is a unique cathode material for LIBs, not only because of its promising high operation voltage and large reversible capacity, but also its amazing crystal structure and related properties which are sensitive to synthesis conditions, and thus providing a good structure-function model for researchers to manipulate and study. Although great advances have been achieved through manipulating its crystal structure, chemical composition, surface chemistry and particle morphologies, still many fundamental issues remain unaddressed, such as low first cycle coulombic efficiency and instability at high potential close to 5 V. For the future, systematic studies and improvements of the cathode, electrolyte and anode are still needed to be able to put $LiNi_{0.5}Mn_{1.5}O_4$ into real life battery applications.

Acknowledgements

This work was supported by the National Natural Science Foundation of China (NSFC 51462029)

References

[1] J.B. Goodenough, Y. Kim, Chemistry of Materials 22 (2010) 587. https://doi.org/10.1021/cm901452z

[2] J.B. Goodenough, Journal of Power Sources 174 (2007) 996. https://doi.org/10.1016/j.jpowsour.2007.06.217

[3] A. Kraytsberg, Y. Ein-Eli, Advanced Energy Materials 2 (2012) 922. https://doi.org/10.1002/aenm.201200068

[4] R. Santhanam, B. Rambabu, Journal of Power Sources 195 (2010) 5442. https://doi.org/10.1016/j.jpowsour.2010.03.067

[5] S.H. Park, S.W. Oh, S.H. Kang, I. Belharouak, K. Amine, Y.K. Sun, Electrochimica Acta 52 (2007) 7226. https://doi.org/10.1016/j.electacta.2007.05.050

[6] J.W. Fergus, Journal of Power Sources 195 (2010) 939. https://doi.org/10.1016/j.jpowsour.2009.08.089

[7] K. Ariyoshi, Y. Iwakoshi, N. Nakayama, T. Ohzuku, Journal of the Electrochemical Society 151 (2004) A296. https://doi.org/10.1149/1.1639162

[8] Q.M. Zhong, A. Bonakdarpour, M.J. Zhang, Y. Gao, J.R. Dahn, Journal of the Electrochemical Society 144 (1997) 205. https://doi.org/10.1149/1.1837386

[9] S.R. Li, C.H. Chen, J. Camardese, J.R. Dahn, Journal of the Electrochemical Society 160 (2013) A1517. https://doi.org/10.1149/2.087309jes

[10] Y.J. Wei, L.Y. Yan, C.Z. Wang, X.G. Xu, F. Wu, G. Chen, The Journal of Physical Chemistry B 108 (2004) 18547. https://doi.org/10.1021/jp0479522

[11] R. Alcantara, M. Jaraba, P. Lavela, J.L. Tirado, E. Zhecheva, R. Stoyanova, Chemistry of Materials 16 (2004) 1573. https://doi.org/10.1021/cm035369c

[12] J. Liu, A. Manthiram, Journal of Physical Chemistry C 113 (2009) 15073. https://doi.org/10.1021/jp904276t

[13] X.L. Wu, S.B. Kim, Journal of Power Sources 109 (2002) 53.
      https://doi.org/10.1016/S0378-7753(02)00034-4

[14] J.H. Kim, S.T. Myung, Y.K. Sun, Electrochimica Acta 49 (2004) 219.
      https://doi.org/10.1016/j.electacta.2003.07.003

[15] J. Lu, Y.-L. Chang, B. Song, H. Xia, J.-R. Yang, K.S. Lee, L. Lu, Journal of
      Power Sources 271 (2014) 604. https://doi.org/10.1016/j.jpowsour.2014.08.037

[16] H. Komatsu, H. Arai, Y. Koyama, K. Sato, T. Kato, R. Yoshida, H. Murayama, I.
      Takahashi, Y. Orikasa, K. Fukuda, T. Hirayama, Y. Ikuhara, Y. Ukyo, Y.
      Uchimoto, Z. Ogumi, Advanced Energy Materials 5 (2015).

[17] M. Kunduraci, J.F. Al-Sharab, G.G. Amatucci, Chemistry of Materials 18 (2006)
      3585. https://doi.org/10.1021/cm060729s

[18] M. Kunduraci, G.G. Amatucci, Electrochimica Acta 53 (2008) 4193.
      https://doi.org/10.1016/j.electacta.2007.12.057

[19] H. Wang, T.A. Tan, P. Yang, M.O. Lai, L. Lu, The Journal of Physical Chemistry
      C 115 (2011) 6102. https://doi.org/10.1021/jp110746w

[20] J.C. Arrebola, A. Caballero, M. Cruz, L. Hernan, J. Morales, E.R. Castellon,
      Advanced Functional Materials 16 (2006) 1904.
      https://doi.org/10.1002/adfm.200500892

[21] J.H. Kim, S.T. Myung, C.S. Yoon, S.G. Kang, Y.K. Sun, Chemistry of Materials
      16 (2004) 906. https://doi.org/10.1021/cm035050s

[22] M. Kunduraci, G.G. Amatucci, Journal of The Electrochemical Society 153 (2006)
      A1345. https://doi.org/10.1149/1.2198110

[23] W.K. Pang, N. Sharma, V.K. Peterson, J.-J. Shiu, S.-h. Wu, Journal of Power
      Sources 246 (2014) 464. https://doi.org/10.1016/j.jpowsour.2013.07.114

[24] L. Wang, H. Li, X. Huang, E. Baudrin, Solid State Ionics 193 (2011) 32.
      https://doi.org/10.1016/j.ssi.2011.04.007

[25] K. Saravanan, A. Jarry, R. Kostecki, G. Chen, Scientific Reports 5 (2015).

[26] H. Wang, Z. Shi, J. Li, S. Yang, R. Ren, J. Cui, J. Xiao, B. Zhang, Journal of
      Power Sources 288 (2015) 206. https://doi.org/10.1016/j.jpowsour.2015.04.137

[27] J. Xiao, X. Chen, P.V. Sushko, M.L. Sushko, L. Kovarik, J. Feng, Z. Deng, J.
      Zheng, G.L. Graff, Z. Nie, D. Choi, J. Liu, J.-G. Zhang, M.S. Whittingham,
      Advanced Materials 24 (2012) 2109. https://doi.org/10.1002/adma.201104767

[28] M. Lin, L. Ben, Y. Sun, H. Wang, Z. Yang, L. Gu, X. Yu, X.-Q. Yang, H. Zhao, R. Yu, M. Armand, X. Huang, Chemistry of Materials 27 (2015) 292. https://doi.org/10.1021/cm503972a

[29] H. Wang, Journal of Nanoscience and Nanotechnology 15 (2015) 6883. https://doi.org/10.1166/jnn.2015.10726

[30] D. Tang, Y. Sun, Z. Yang, L. Ben, L. Gu, X. Huang, Chemistry of Materials 26 (2014) 3535. https://doi.org/10.1021/cm501125e

[31] M.R. Palacin, Y. Chabre, L. Dupont, M. Hervieu, P. Strobel, G. Rousse, C. Masquelier, M. Anne, G.G. Amatucci, J.M. Tarascon, Journal of the Electrochemical Society 147 (2000) 845. https://doi.org/10.1149/1.1393281

[32] I. Takahashi, H. Arai, H. Murayama, K. Sato, H. Komatsu, H. Tanida, Y. Koyama, Y. Uchimoto, Z. Ogumi, Physical Chemistry Chemical Physics 18 (2016) 1897. https://doi.org/10.1039/C5CP05535K

[33] Y. Chen, Y. Sun, X. Huang, Computational Materials Science 115 (2016) 109. https://doi.org/10.1016/j.commatsci.2016.01.005

[34] Y. Zhang, Y. Li, X. Xia, X. Wang, C. Gu, J. Tu, Science China-Technological Sciences 58 (2015) 1809. https://doi.org/10.1007/s11431-015-5933-x

[35] Y. Xue, Z. Wang, L. Zheng, F. Yu, B. Liu, Y. Zhang, K. Ke, Scientific Reports 5 (2015).

[36] Z. Wang, Q. Su, H. Deng, Y. Fu, Chemelectrochem 2 (2015) 1182. https://doi.org/10.1002/celc.201500059

[37] Q. Wu, Y. Liu, C.S. Johnson, Y. Li, D.W. Dees, W. Lu, Chemistry of Materials (2014).

[38] G. Liu, K.-S. Park, J. Song, J.B. Goodenough, Journal of Power Sources 243 (2013) 260. https://doi.org/10.1016/j.jpowsour.2013.05.189

[39] E. Hu, S.-M. Bak, J. Liu, X. Yu, Y. Zhou, S.N. Ehrlich, X.-Q. Yang, K.-W. Nam, Chemistry of Materials 26 (2013) 1108. https://doi.org/10.1021/cm403400y

[40] J. Zheng, J. Xiao, X. Yu, L. Kovarik, M. Gu, F. Omenya, X. Chen, X.-Q. Yang, J. Liu, G.L. Graff, M.S. Whittingham, J.-G. Zhang, Physical Chemistry Chemical Physics 14 (2012) 13515. https://doi.org/10.1039/c2cp43007j

[41] R. Qiao, Y. Wang, P. Olalde-Velasco, H. Li, Y.-S. Hu, W. Yang, Journal of Power Sources 273 (2015) 1120. https://doi.org/10.1016/j.jpowsour.2014.10.013

[42] N.S. Norberg, S.F. Lux, R. Kostecki, Electrochemistry Communications 34 (2013) 29. https://doi.org/10.1016/j.elecom.2013.04.007

[43] M. Kunduraci, G.G. Amatucci, Journal of Power Sources 165 (2007) 359. https://doi.org/10.1016/j.jpowsour.2006.11.051

[44] Z. Moorhead-Rosenberg, A. Huq, J.B. Goodenough, A. Manthiram, Chemistry of Materials 27 (2015) 6934. https://doi.org/10.1021/acs.chemmater.5b01356

[45] J. Lee, C. Kim, B. Kang, Npg Asia Materials 7 (2015).

[46] B. Leon, J.M. Lloris, C.P. Vicente, J.L. Tirado, Electrochemical and Solid State Letters 9 (2006) A96. https://doi.org/10.1149/1.2154374

[47] S.B. Park, W.S. Eom, W.I. Cho, H. Jang, Journal of Power Sources 159 (2006) 679. https://doi.org/10.1016/j.jpowsour.2005.10.099

[48] G.B. Zhong, Y.Y. Wang, Z.C. Zhang, C.H. Chen, Electrochimica Acta 56 (2011) 6554. https://doi.org/10.1016/j.electacta.2011.03.093

[49] J. Liu, Z. Sun, J. Xie, H. Chen, N. Wu, B. Wu, Journal of Power Sources 240 (2013) 95. https://doi.org/10.1016/j.jpowsour.2013.03.172

[50] S.H. Oh, K.Y. Chung, S.H. Jeon, C.S. Kim, W.I. Cho, B.W. Cho, Journal of Alloys and Compounds 469 (2009) 244. https://doi.org/10.1016/j.jallcom.2008.01.097

[51] M.-C. Yang, B. Xu, J.-H. Cheng, C.-J. Pan, B.-J. Hwang, Y.S. Meng, Chemistry of Materials 23 (2011) 2832. https://doi.org/10.1021/cm200042z

[52] Y. Ein-Eli, J.T. Vaughey, M.M. Thackeray, S. Mukerjee, X.Q. Yang, J. McBreen, Journal of the Electrochemical Society 146 (1999) 908. https://doi.org/10.1149/1.1391699

[53] G.B. Zhong, Y.Y. Wang, X.J. Zhao, Q.S. Wang, Y. Yu, C.H. Chen, Journal of Power Sources 216 (2012) 368. https://doi.org/10.1016/j.jpowsour.2012.05.108

[54] G.B. Zhong, Y.Y. Wang, Y.Q. Yu, C.H. Chen, Journal of Power Sources 205 (2012) 385. https://doi.org/10.1016/j.jpowsour.2011.12.037

[55] G.Q. Liu, L. Wen, G.Y. Liu, Y.W. Tian, Journal of Alloys and Compounds 501 (2010) 233. https://doi.org/10.1016/j.jallcom.2010.04.076

[56] M. Aklalouch, R.M. Rojas, J.M. Rojo, I. Saadoune, J.M. Amarilla, Electrochimica Acta 54 (2009) 7542. https://doi.org/10.1016/j.electacta.2009.08.012

[57]   T.F. Yi, C.Y. Li, Y.R. Zhu, J. Shu, R.S. Zhu, Journal of Solid State
         Electrochemistry 13 (2009) 913. https://doi.org/10.1007/s10008-008-0628-x

[58]   A. Ito, D. Li, Y. Lee, K. Kobayakawa, Y. Sato, Journal of Power Sources 185
         (2008) 1429. https://doi.org/10.1016/j.jpowsour.2008.08.087

[59]   D.C. Li, A. Ito, K. Kobayakawa, H. Noguchi, Y. Sato, Journal of Power Sources
         161 (2006) 1241. https://doi.org/10.1016/j.jpowsour.2006.04.120

[60]   P. Sun, Y. Ma, T. Zhai, H. Li, Electrochimica Acta 191 (2016) 237.
         https://doi.org/10.1016/j.electacta.2016.01.087

[61]   G.H. Lee, H.S. Kim, S.G. Baek, H.J. Choi, K.Y. Chung, B.W. Cho, S.Y. Lee, Y.-
         S. Lee, Journal of Power Sources 298 (2015) 379.
         https://doi.org/10.1016/j.jpowsour.2015.08.053

[62]   M. Mo, K.S. Hui, X. Hong, J. Guo, C. Ye, A. Li, N. Hu, Z. Huang, J. Jiang, J.
         Liang, H. Chen, Applied Surface Science 290 (2014) 412.
         https://doi.org/10.1016/j.apsusc.2013.11.094

[63]   W. Zhu, D. Liu, J. Trottier, C. Gagnon, A. Mauger, C.M. Julien, K. Zaghib,
         Journal of Power Sources 242 (2013) 236.
         https://doi.org/10.1016/j.jpowsour.2013.05.021

[64]   E.-S. Lee, A. Manthiram, Journal of Materials Chemistry A 1 (2013) 3118.
         https://doi.org/10.1039/c2ta01171a

[65]   Y.-P. Zeng, X.-l. Wu, P. Mei, L.-N. Cong, C. Yao, R.-S. Wang, H.-M. Xie, L.-Q.
         Sun, Electrochimica Acta 138 (2014) 493.
         https://doi.org/10.1016/j.electacta.2014.06.082

[66]   N.M. Hagh, G.G. Amatucci, Journal of Power Sources 256 (2014) 457.
         https://doi.org/10.1016/j.jpowsour.2013.12.135

[67]   H. Li, Y. Luo, J. Xie, Q. Zhang, L. Yan, Journal of Alloys and Compounds 639
         (2015) 346. https://doi.org/10.1016/j.jallcom.2015.03.114

[68]   G.-D. Du, Y.-N. Nuli, Z.-Z. Feng, J.-L. Wang, J. Yang, Acta Physico-Chimica
         Sinica 24 (2008) 165.

[69]   M. Wagemaker, F.G.B. Ooms, E.M. Kelder, J. Schoonman, G.J. Kearley, F.M.
         Mulder, Journal of the American Chemical Society 126 (2004) 13526.
         https://doi.org/10.1021/ja048319x

[70]   T.A. Arunkumar, A. Manthiram, Electrochemical and Solid-State Letters 8 (2005)
         A403. https://doi.org/10.1149/1.1945369

[71]   M. Aklalouch, J.M. Amarilla, R.M. Rojas, I. Saadoune, J.M. Rojo, Electrochemistry Communications 12 (2010) 548. https://doi.org/10.1016/j.elecom.2010.01.040

[72]   J. Liu, A. Manthiram, Chemistry of Materials 21 (2009) 1695. https://doi.org/10.1021/cm9000043

[73]   H. Wang, J. Li, S. Yang, B. Zhang, J. Xiao, R. Ren, J. Cui, W. Xiao, Materials Technology 30 (2015) A75. https://doi.org/10.1179/17535557A15Y.000000008

[74]   K. Xu, A. von Cresce, Journal of Materials Chemistry 21 (2011) 9849. https://doi.org/10.1039/c0jm04309e

[75]   Y.K. Sun, C.S. Yoon, I.H. Oh, Electrochimica Acta 48 (2003) 503. https://doi.org/10.1016/S0013-4686(02)00717-X

[76]   J.C. Arrebola, A. Caballero, L. Hernan, J. Morales, Electrochemical and Solid State Letters 8 (2005) A641. https://doi.org/10.1149/1.2116147

[77]   T. Noguchi, I. Yamazaki, T. Numata, M. Shirakata, Journal of Power Sources 174 (2007) 359. https://doi.org/10.1016/j.jpowsour.2007.06.139

[78]   Y.K. Fan, J.M. Wang, Z. Tang, W.C. He, J.Q. Zhang, Electrochimica Acta 52 (2007) 3870. https://doi.org/10.1016/j.electacta.2006.10.063

[79]   H.M. Wu, I. Belharouak, A. Abouimrane, Y.K. Sun, K. Amine, Journal of Power Sources 195 (2010) 2909. https://doi.org/10.1016/j.jpowsour.2009.11.029

[80]   J. Chong, S. Xun, X. Song, G. Liu, V.S. Battaglia, Nano Energy 2 (2013) 283. https://doi.org/10.1016/j.nanoen.2012.09.013

[81]   K.W. Leitner, H. Wolf, A. Garsuch, F. Chesneau, M. Schulz-Dobrick, Journal of Power Sources 244 (2013) 548. https://doi.org/10.1016/j.jpowsour.2013.01.187

[82]   L. Baggetto, N.J. Dudney, G.M. Veith, Electrochimica Acta 90 (2013) 135. https://doi.org/10.1016/j.electacta.2012.11.120

[83]   J.C. Arrebola, A. Caballero, L. Hernán, J. Morales, Journal of Power Sources 195 (2010) 4278. https://doi.org/10.1016/j.jpowsour.2010.01.004

[84]   Y.K. Sun, K.J. Hong, J. Prakash, K. Amine, Electrochemistry Communications 4 (2002) 344. https://doi.org/10.1016/S1388-2481(02)00277-1

[85]   D. Liu, Y. Bai, S. Zhao, W. Zhang, Journal of Power Sources 219 (2012) 333. https://doi.org/10.1016/j.jpowsour.2012.07.058

[86]   J.Y. Shi, C.-W. Yi, K. Kim, Journal of Power Sources 195 (2010) 6860. https://doi.org/10.1016/j.jpowsour.2010.02.063

[87]   J. Arrebola, A. Caballero, L. Hernán, J. Morales, E. Rodríguez Castellón, J.R. Ramos Barrado, Journal of The Electrochemical Society 154 (2007) A178. https://doi.org/10.1149/1.2426799

[88]   Y.-F. Deng, S.-X. Zhao, Y.-H. Xu, C.-W. Nan, Journal of Power Sources 296 (2015) 261. https://doi.org/10.1016/j.jpowsour.2015.07.017

[89]   J. Chong, J. Zhang, H. Xie, X. Song, G. Liu, V. Battaglia, S. Xun, R. Wang, Rsc Advances 6 (2016) 19245. https://doi.org/10.1039/C6RA00119J

[90]   S. Niketic, M. Couillard, D. MacNeil, Y. Abu-Lebdeh, Journal of Power Sources 271 (2014) 285. https://doi.org/10.1016/j.jpowsour.2014.08.015

[91]   H. Li, H. Zhou, Chemical Communications 48 (2012) 1201. https://doi.org/10.1039/C1CC14764A

[92]   Y.K. Sun, Y.S. Lee, M. Yoshio, K. Amine, Journal of the Electrochemical Society 150 (2003) L11. https://doi.org/10.1149/1.1566967

[93]   Y. Kobayashi, H. Miyashiro, K. Takei, H. Shigemura, M. Tabuchi, H. Kageyama, T. Iwahori, Journal of the Electrochemical Society 150 (2003) A1577. https://doi.org/10.1149/1.1619988

[94]   H. Rong, M. Xu, B. Xie, X. Liao, W. Huang, L. Xing, W. Li, Electrochimica Acta 147 (2014) 31. https://doi.org/10.1016/j.electacta.2014.09.105

[95]   J. Liu, A. Manthiram, Journal of the Electrochemical Society 156 (2009) A66. https://doi.org/10.1149/1.3028318

[96]   C.A. Kim, H.J. Choi, J.H. Lee, S.Y. Yoo, J.W. Kim, J.H. Shim, B. Kang, Electrochimica Acta 184 (2015) 134. https://doi.org/10.1016/j.electacta.2015.10.041

[97]   H.-M. Cho, M.V. Chen, A.C. MacRae, Y.S. Meng, Acs Applied Materials & Interfaces 7 (2015) 16231. https://doi.org/10.1021/acsami.5b01392

[98]   J. Arrebola, A. Caballero, L. Hernán, J. Morales, E. Rodríguez Castellón, Electrochemical and Solid-State Letters 8 (2005) A303. https://doi.org/10.1149/1.1911877

[99]   S. Lee, Y. Cho, H.-K. Song, K.T. Lee, J. Cho, Angewandte Chemie International Edition 51 (2012) 8748. https://doi.org/10.1002/anie.201203581

[100] H. Wang, M. Yoshio, T. Abe, Z. Ogumi, Journal of The Electrochemical Society 149 (2002) A499. https://doi.org/10.1149/1.1461377

[101] N. Zhang, T. Yang, Y. Lang, K. Sun, Journal of Alloys and Compounds 509 (2011) 3783. https://doi.org/10.1016/j.jallcom.2010.12.188

[102] H. Xia, Z. Luo, J. Xie, Progress in Natural Science: Materials International 22 (2012) 572.

[103] T. Yang, N. Zhang, Y. Lang, K. Sun, Electrochimica Acta 56 (2011) 4058. https://doi.org/10.1016/j.electacta.2010.12.109

[104] L. Xue, Y. Liao, L. Yang, X. Li, W. Li, Ionics 21 (2015) 1269. https://doi.org/10.1007/s11581-014-1286-x

[105] Y.-Z. Jin, Y.-Z. Lv, Y. Xue, J. Wu, X.-G. Zhang, Z.-B. Wang, Rsc Advances 4 (2014) 57041. https://doi.org/10.1039/C4RA07921C

[106] J. Liu, H. Xu, X. Jiang, J. Yang, Y. Qian, Journal of Power Sources 231 (2013) 39. https://doi.org/10.1016/j.jpowsour.2012.12.071

[107] A. Vu, A. Stein, Chemistry of Materials 23 (2011) 3237. https://doi.org/10.1021/cm201197j

[108] J.-H. Cho, J.-H. Park, M.-H. Lee, H.-K. Song, S.-Y. Lee, Energy & Environmental Science 5 (2012) 7124. https://doi.org/10.1039/c2ee03389e

[109] X. Fang, M. Ge, J. Rong, C. Zhou, Journal of Materials Chemistry A 1 (2013) 4083. https://doi.org/10.1039/c3ta01534c

[110] X. Tang, S.S. Jan, Y. Qian, H. Xia, J. Ni, S.V. Savilov, S.M. Aldoshin, Scientific Reports 5 (2015).

[111] C.R. Sides, N.C. Li, C.J. Patrissi, B. Scrosati, C.R. Martin, Mrs Bulletin 27 (2002) 604. https://doi.org/10.1557/mrs2002.195

[112] A. Odani, A. Nimberger, B. Markovsky, E. Sominski, E. Levi, V.G. Kumar, A. Motiei, A. Gedanken, P. Dan, D. Aurbach, Journal of Power Sources 119 (2003) 517. https://doi.org/10.1016/S0378-7753(03)00276-3

[113] B. Ellis, P.S. Herle, Y.H. Rho, L.F. Nazar, R. Dunlap, L.K. Perry, D.H. Ryan, Faraday Discussions 134 (2007) 119. https://doi.org/10.1039/B602698B

[114] K.M. Shaju, P.G. Bruce, Dalton Transactions (2008) 5471. https://doi.org/10.1039/b806662k

[115] Y. Talyosef, B. Markovsky, R. Lavi, G. Salitra, D. Aurbach, D. Kovacheva, M. Gorova, E. Zhecheva, R. Stoyanova, Journal of the Electrochemical Society 154 (2007) A682. https://doi.org/10.1149/1.2736657

[116] P.G. Bruce, B. Scrosati, J.-M. Tarascon, Angewandte Chemie International Edition 47 (2008) 2930. https://doi.org/10.1002/anie.200702505

[117] Y.S. Lee, Y.K. Sun, S. Ota, T. Miyashita, M. Yoshi, Electrochemistry Communications 4 (2002) 989. https://doi.org/10.1016/S1388-2481(02)00491-5

[118] S.T. Myung, S. Komaba, N. Kumagai, H. Yashiro, H.T. Chung, T.H. Cho, Electrochimica Acta 47 (2002) 2543. https://doi.org/10.1016/S0013-4686(02)00131-7

[119] X. Hao, M.H. Austin, B.M. Bartlett, Dalton Transactions 41 (2012) 8067. https://doi.org/10.1039/c2dt30351e

[120] L.F. Xiao, Y.Q. Zhao, Y.Y. Yang, X.P. Ai, H.X. Yang, Y.L. Cao, Journal of Solid State Electrochemistry 12 (2008) 687. https://doi.org/10.1007/s10008-007-0409-y

[121] W. Luo, Journal of Alloys and Compounds 636 (2015) 24. https://doi.org/10.1016/j.jallcom.2015.02.163

[122] L. Xue, X. Li, Y. Liao, L. Xing, M. Xu, W. Li, Journal of Solid State Electrochemistry 19 (2015) 569. https://doi.org/10.1007/s10008-014-2635-4

[123] M.A. Kiani, M.S. Rahmanifar, M.F. El-Kady, R.B. Kaner, M.F. Mousavi, Rsc Advances 5 (2015) 50433. https://doi.org/10.1039/C5RA08170J

[124] Y.-C. Jin, M.-I. Lu, T.-H. Wang, C.-R. Yang, J.-G. Duh, Journal of Power Sources 262 (2014) 483. https://doi.org/10.1016/j.jpowsour.2014.03.089

[125] S. Ullah, F. Ahmed, A. Badshah, A.A. Altaf, R. Raza, B. Lal, R. Hussain, Australian Journal of Chemistry 67 (2014) 289.

[126] J. Yang, X. Han, X. Zhang, F. Cheng, J. Chen, Nano Research 6 (2013) 679. https://doi.org/10.1007/s12274-013-0343-5

[127] Y.-C. Jin, J.-G. Duh, Materials Letters 93 (2013) 77. https://doi.org/10.1016/j.matlet.2012.11.039

[128] H.-M. Cho, Y.S. Meng, Journal of the Electrochemical Society 160 (2013) A1482. https://doi.org/10.1149/2.059309jes

[129] X. Zhang, F. Cheng, K. Zhang, Y. Liang, S. Yang, J. Liang, J. Chen, Rsc Advances 2 (2012) 5669. https://doi.org/10.1039/c2ra20669b

[130] X. Huang, Q. Zhang, J. Gan, H. Chang, Y. Yang, Journal of the Electrochemical Society 158 (2011) A139. https://doi.org/10.1149/1.3521292

[131] X. Fang, Y. Lu, N. Ding, X.Y. Feng, C. Liu, C.H. Chen, Electrochimica Acta 55 (2010) 832. https://doi.org/10.1016/j.electacta.2009.09.046

[132] L. Xiao, Y. Zhao, Y. Yang, X. Ai, H. Yang, Y. Cao, Journal of Solid State Electrochemistry 12 (2008) 687. https://doi.org/10.1007/s10008-007-0409-y

[133] H.S. Fang, L.P. Li, G.S. Li, Journal of Power Sources 167 (2007) 223. https://doi.org/10.1016/j.jpowsour.2007.02.015

[134] M.G. Lazarraga, L. Pascual, H. Gadjov, D. Kovacheva, K. Petrov, J.M. Amarilla, R.M. Rojas, M.A. Martin-Luengo, J.M. Rojo, Journal of Materials Chemistry 14 (2004) 1640. https://doi.org/10.1039/b314157h

[135] X. Zhang, F. Cheng, J. Yang, J. Chen, Nano Letters 13 (2013) 2822. https://doi.org/10.1021/nl401072x

[136] Y. Liu, M. Zhang, Y. Xia, B. Qiu, Z. Liu, X. Li, Journal of Power Sources 256 (2014) 66. https://doi.org/10.1016/j.jpowsour.2014.01.059

[137] Z. Chen, S. Qiu, Y. Cao, X. Ai, K. Xie, X. Hong, H. Yang, Journal of Materials Chemistry 22 (2012) 17768. https://doi.org/10.1039/c2jm33338d

CHAPTER 3

Monoclinic $Li_3V_2(PO_4)_3$ and Its Derivatives as Cathode Materials for Lithium-ion Batteries

Quanqi Chen*, Xu Yan, Xinmei Zhang

College of Chemistry and Bioengineering, Guilin University of Technology

Guilin 541004, P.R. China

* quanqi.chen@glut.edu.cn; quanqi.chen@yahoo.com

Abstract

As a polyanion compound with PO_4^{3-}, monoclinic lithium vanadium phosphate $(Li_3V_2(PO_4)_3)$ possesses excellent structural and thermal stability and exhibits high capacity (197 mAh/g), high operating voltage (about 3.8 V) and better cycle performance, thus $Li_3V_2(PO_4)_3$ has been considered as a promising cathode material for lithium-ion batteries. The physical and electrochemical properties, synthesis methods, and development of $Li_3V_2(PO_4)_3$ and its derivatives are also introduced in detail in this chapter.

Keywords

Lithium Vanadium Phosphate, Monoclinic, Polyanion Cathode, Thermal Stability

Contents

1.   Introduction

Recently, monoclinic lithium vanadium phosphate ($Li_3V_2(PO_4)_3$) and its derivatives have been considered as promising cathode materials for lithium-ion batteries because of their advantages such as high capacity and energy density, better cyclability and excellent thermal stability. The theoretical capacity of monoclinic $Li_3V_2(PO_4)_3$ is as high as 197 $mAh \cdot g^{-1}$ based on the assumption that three moles of Li^+ can be completely and reversibly extracted from per mole of $Li_3V_2(PO_4)_3$ or inserted into per mole of delithiated $V_2(PO_4)_3$ [1]. The monoclinic $Li_3V_2(PO_4)_3$ can deliver average operating voltage about 3.8 V (vs. Li^+/Li)[2] and high capacity, therefore the high energy and large density of $Li_3V_2(PO_4)_3$ result in its high energy density. As a polyanion compound, $Li_3V_2(PO_4)_3$ is also of excellent structural stability and thermal stability. This chapter gives a detailed introduction of monoclinic $Li_3V_2(PO_4)_3$ and the latest research advances in this area.

2.   Basics of Lithium Vanadium Phosphate

2.1   Structure of Lithium Vanadium Phosphate

Lithium vanadium phosphates exhibit a variety of structures based on the connectivity of PO_4 tetrahedra and VO_6 octahedra, which can accommodate Li^+ ions in interstitial sites within the framework. Materials of the stoichiometric $Li_3V_2(PO_4)_3$ are of two different frameworks, rhombohedral phase and monoclinic phase. The crystal structure of rhombohedral $Li_3V_2(PO_4)_3$ belongs to space group $R\bar{3}$ and its cell parameters are as follows: a=8.316 Å, b=22.484 Å [3], and the corresponding framework is depicted in Fig. 1A, isotopic with those of $Li_3Fe_2(PO_4)_3$, $Li_3Ti_2(PO_4)_3$, $Li_3Cr_2(PO_4)_3$ and $Li_3In_2(PO_4)_3$. It can be seen that the framework consists of VO_6 octahedra and PO_4 tetrahedra linked with their vertices, forming $V_2(PO_4)_3$ "lantern" units stacked along the [001] direction. Lithium is situated on a unique 4-fold coordinated crystallographic site [4]. While the crystal structure of monoclinic $Li_3V_2(PO_4)_3$ is assigned to space group $P2_1/n$ and the cell parameters are listed as follows: a=8.605 Å, b=8.591 Å, c=12.038 Å, β=90.60°. Fig.1B indicates that monoclinic framework comprises of slightly distorted PO_4 tetrahedra and VO_6 octahedra interconnected by common apical oxygens to form a $(V-O-P)_n$ bonding arrangement[4]. Its $V_2(PO_4)_3$ "lantern" units are alternately oriented perpendicular to each other, which create a closer packing than those in rhombohedral $Li_3V_2(PO_4)_3$, thus minimizing the free volume of the interstitial space. It contains two vanadium sites $V(1)$ and $V(2)$ with average V-O bond length of 2.003 and 2.006 Å, respectively. Three distinct lithium atoms occupy three distinct crystallographic positions in the interstitial voids, $Li(1)$ is surrounded by four oxygen atoms to form a tetrahedron, and $Li(2)$ and $Li(3)$ occupy highly distorted tetrahedral environments that can be better described as five-coordinate Li-O sites, where the fifth Li-O bond is long as 2.6 Å. The local Li environments are also distinguishable by ^7Li MAS NMR. The spectrum of monoclinic $Li_3V_2(PO_4)_3$ displays three well-resolved isotropic chemical shifts at 103, 52, and 17 ppm, which correspond to the three crystallographic Li sites, respectively [1]. The presence of corner-shared chains of Li polyhedra along the a-axis and open diffusion pathways in the other directions should lead to rapid, isotropic ionic transport similar to the fast-ion conducting NASICON phase. The mobility of these three lithium sites, as determined by two-dimensional exchange spectroscopy, occurs on the microsecond time scale, which is faster than the millisecond Li hopping processes in spinel $LiMn_2O_4$. On the other hand, monoclinic $Li_3V_2(PO_4)_3$ undergoes three structural changes in the sequence of α-phase (monoclinic), β-phase (orthorhombic) and γ-phase (orthorhombic) between room temperature and 573 K, which is confirmed by differential scanning calorimetry (DSC), temperature-controlled X-ray diffraction (XRD) and Raman spectra [5]. These phase transitions are reversible upon cooling back to room temperature. It is noted that monoclinic $Li_3V_2(PO_4)_3$ is not stable and is oxidized to high-valent vanadium oxide in air at temperature above 800 K.

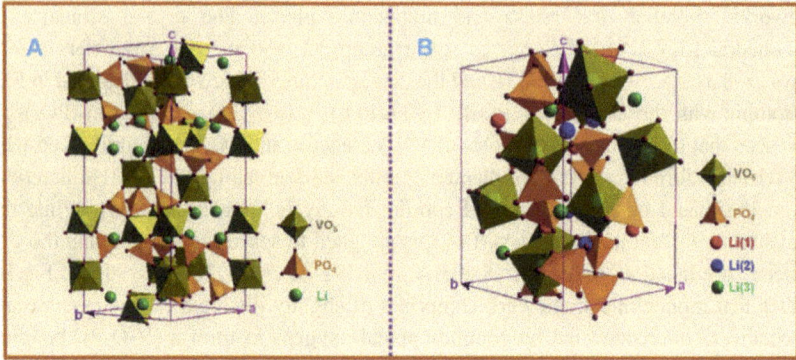

Figure 1. Crystal structure of rhombohedral (A) and monoclinic (B) $Li_3V_2(PO_4)_3$. Reproduced with permission from Ref. [6].

2.2   Mechanism of Lithium Insertion and Extraction

The voltage-composition profiles of rhombohedral $Li_3V_2(PO_4)_3$ are significantly different from those of monoclinic $Li_3V_2(PO_4)_3$, which results from the structural differences between rhombohedral and monoclinic $Li_3V_2(PO_4)_3$. In theory, 2 Li can be reversibly extracted from per rhombohedral $Li_3V_2(PO_4)_3$ unit and consequently $LiV_2(PO_4)_3$ is formed. At the same time, 2 Li can be reversibly inserted into per delithiated $LiV_2(PO_4)_3$ unit and the theoretical capacity of rhombohedral $Li_3V_2(PO_4)_3$ is about 133 mAh/g. On the basis of previous reports, the electrochemical reactions between $Li_3V_2(PO_4)_3$ and $LiV_2(PO_4)_3$ is a two-phase transition, which is associated with the redox reactions between V^{3+} and V^{4+}. The rhombohedral $Li_3V_2(PO_4)_3$ shows one charge voltage plateau and one discharge voltage plateau and both two plateaus are located at about 3.7 V (vs. Li^+/Li), corresponding to V^{3+}/V^{4+} redox couple. Unlike to rhombohedral $Li_3V_2(PO_4)_3$, the monoclinic phase can provide three reversible extraction/insertion Li^+ and hence has a high theoretical capacity of 197 mAh/g and better electrochemical performance than rhombohedral $Li_3V_2(PO_4)_3$. Therefore, this chapter focuses on the monoclinic phase, and the following $Li_3V_2(PO_4)_3$ refers to monoclinic $Li_3V_2(PO_4)_3$ unless specified otherwise.

In the voltage range of 3.0-4.3 V (vs. Li^+/Li), 2 Li can be theoretically extracted from per $Li_3V_2(PO_4)_3$ unit and the corresponding charge profile displays three voltage plateaus situating at 3.61, 3.69 and 4.06 V (vs. Li^+/Li) as shown in Fig.2a, respectively, which relates with three two-phase transition reactions [2]. The matched reactions are listed in sequence as follows:

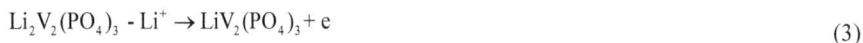

$$Li_3V_2(PO_4)_3 - 0.5Li^+ \rightarrow Li_{2.5}V_2(PO_4)_3 + 0.5e \tag{1}$$

$$Li_{2.5}V_2(PO_4)_3 - 0.5Li^+ \rightarrow Li_2V_2(PO_4)_3 + 0.5e \tag{2}$$

$$Li_2V_2(PO_4)_3 - Li^+ \rightarrow LiV_2(PO_4)_3 + e \tag{3}$$

When $Li_3V_2(PO_4)_3$ is charged, Li^+ is firstly extracted from Li(3) because this position has the highest energy among all the three sites, Li(1), Li(2) and Li(3). At the same time, Li^+ residing in Li(2) moves to a tetrahedral site with similar energy to the Li(1) site. The Li^+ extraction from Li(3) includes two steps because of the existence of an intermediate structure, $Li_{2.5}V_2(PO_4)_3$. One Li^+ extraction results in the valence change of one vanadium from +3 to +4. Subsequent one Li^+ extraction from Li(1) turns another V^{3+} to V^{4+}. After extraction of two Li^+ from per $Li_3V_2(PO_4)_3$ unit, the volume changes from 899.8 Å^3 for $Li_3V_2(PO_4)_3$ to 823.8 Å^3 for monoclinic $LiV_2(PO_4)_3$, corresponding to volume contraction of about 8.4% [1], which suggests that such a small structural deformation is beneficial for better cycle performance during charge and discharge processes. Similar to the charge profile, discharge profile also presents three voltage profiles at about 4.03, 3.64 and 3.55 V (vs. Li^+/Li), respectively, and the corresponding three two-phase transition reactions are written as the following electrochemical equations:

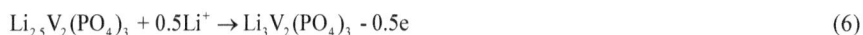

$$LiV_2(PO_4)_3 + Li^+ \rightarrow Li_2V_2(PO_4)_3 - e \tag{4}$$

$$Li_2V_2(PO_4)_3 + 0.5Li^+ \rightarrow Li_{2.5}V_2(PO_4)_3 - 0.5e \tag{5}$$

$$Li_{2.5}V_2(PO_4)_3 + 0.5Li^+ \rightarrow Li_3V_2(PO_4)_3 - 0.5e \tag{6}$$

Upon discharge, one Li^+ is intercalated into $LiV_2(PO_4)_3$ forming $Li_2V_2(PO_4)_3$. As a result, one V^{4+} is reduced to V^{3+}. Subsequently another Li^+ is inserted into $Li_2V_2(PO_4)_3$ and this intercalation involves two steps, which leads to reduction of another V^{4+} to V^{3+} and $Li_3V_2(PO_4)_3$ is formed.

In the voltage range of 3.0-4.8 V (vs. Li^+/Li), 3 Li can be de-inserted from per $Li_3V_2(PO_4)_3$ unit and fully de-lithiated $V_2(PO_4)_3$ is formed in theory. The charge profile exhibits four voltage plateaus including the above-mentioned three plateaus and another one located at about 4.55 V and the fourth two-phase transition reaction is depicted as:

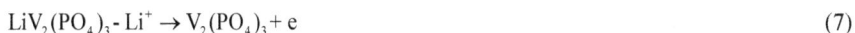

$$LiV_2(PO_4)_3 - Li^+ \rightarrow V_2(PO_4)_3 + e \tag{7}$$

While the discharge profile presented in Fig.2b does not display the fourth discharge plateau, indicating that the transition of $V_2(PO_4)_3$ to $Li_2V_2(PO_4)_3$ is a solid solution behavior, and the corresponding electrochemical reaction is as follows:

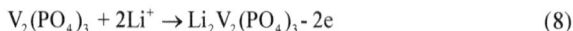

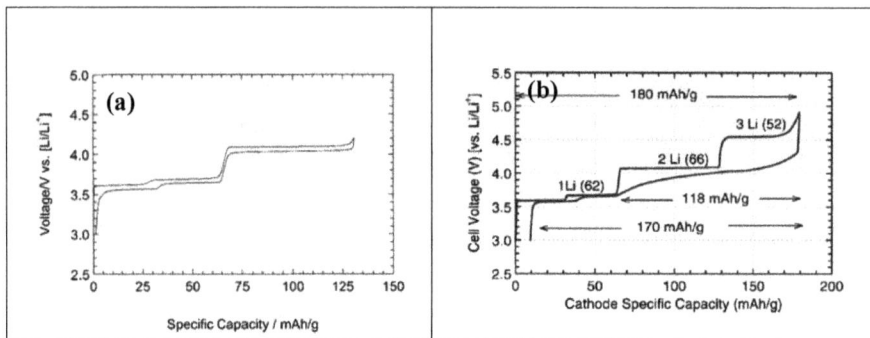

$$V_2(PO_4)_3 + 2Li^+ \rightarrow Li_2V_2(PO_4)_3 - 2e \qquad (8)$$

Figure 2. The charge/discharge profiles of $Li_3V_2(PO_4)_3$ in the voltage range of (a)3.0-4.3, (b)3.0-4.8 V (vs. Li^+/Li). Reproduced with permission from Ref. [7].

2.3   Synthesis

It is well known that kinetics of Li^+ extraction/insertion is a crucial factor to influence the electrochemical performance, such as capacity, rate capability and cycle life, of electroactive materials for lithium ion batteries and is mainly dependent on the structure, electronic and ionic conductivities of electroactive materials. Therefore, the proposed strategies and synthesis methods to improve electrochemical performance of $Li_3V_2(PO_4)_3$ should be based on the advantages and disadvantages of $Li_3V_2(PO_4)_3$. As a polyanion compound with PO_4^{3-}, $Li_3V_2(PO_4)_3$ possesses excellent structural and thermal stability, good Li^+ transport, and high operating voltage resulting from inductive effect of PO_4^{3-}. However, the electronically insulating PO_4^{3-} also isolates the valence electrons of vanadium within the lattice, leading to low intrinsic electronic conductivity. The pristine $Li_3V_2(PO_4)_3$ exhibits a negligible electronic conductivity (2×10^{-8} S/cm [1], 7.73×10^{-8} S/cm [8]), although higher than that of $LiFePO_4$ (~ 10^{-9} S/cm) , unfortunately much lower than that of $LiCoO_2$ (~10^{-3} S/cm) [9] and of $LiMn_2O_4$ (~10^{-4} S/cm) [11]. The low electronic conductivity results in sluggish kinetics of Li^+ extraction/insertion and hence the capacity of pure $Li_3V_2(PO_4)_3$ is only partially accessible even at slow rates, showing a fatal drawback to practical applications of $Li_3V_2(PO_4)_3$. Thus the best strategy to improve the electrochemical performance of $Li_3V_2(PO_4)_3$ is to enhance its electronic conductivity. Methods such as surface modification with conductive materials including carbon, conductive polymer and metals, and metal doping, or a combination of them have been proved to remarkably elevate the electronic conductivity of $Li_3V_2(PO_4)_3$ and consequently improve the electrochemical performance of $Li_3V_2(PO_4)_3$. Summary of the previous literatures about $Li_3V_2(PO_4)_3$ indicates that a majority of synthesis methods are established on the carbothermal reduction method and the resulted products usually contain residual carbon. The specific synthesis methods are described as follows.

2.3.1 Solid State Reaction

Solid state reaction is the most common method to prepare ceramics and is simple and easy to be applied in industry. This synthesis method always consists of mixing solid raw materials by hand grinding or ball milling, pelletizing the resulted mixture, and sintering the pelleted mixture in sequence. In the case of $Li_3V_2(PO_4)_3$, the raw materials commonly include vanadium (*e.g.* V_2O_5, NH_4VO_3 and V_2O_3), lithium (*e.g.* $LiOH$, Li_2CO_3, CH_3COOLi and LiF), phosphorous (*e.g.* $NH_4H_2PO_4$ and $(NH_4)_2HPO_4$), and carbon (*e.g.* acetylene black, activated carbon, and organic compounds) sources. After complete mixing, the mixture was initially heated to 300-400 $^\circ$C to expel gases (*e.g.* NH_3, H_2O, and volatile organic compounds), and then reground, pelleted, and calcined at a temperature ranging from 600 to 1000 $^\circ$C for 10-24 h under inert gas or slightly reducing atmosphere. During the preparation process, the pelletizing is favorable for intimate contact between reactants and hence the complete reaction. During this synthesis, carbon directly added or in situ formed from pyrolysis of the added organic precursor plays multiple functions in the improvement of electrochemical performance of $Li_3V_2(PO_4)_3$. Carbon can act as a reductant to reduce V^{5+} to V^{3+}, and an inhibitor to restrain growth and agglomeration of $Li_3V_2(PO_4)_3$ crystals during calcination in high temperature. Furthermore, the residual carbon can significantly enhance the electronic conductivity. The above-mentioned benefits of carbon can remarkably improve the electrochemical performance of $Li_3V_2(PO_4)_3$. It is well known that Gibbs energy of reaction is closely related to reaction conditions, Gibbs energy of products and reactants, and reaction conditions significantly influence the properties of products. Therefore, properties of $Li_3V_2(PO_4)_3$ are intimately linked with the reactants, calcination temperature and time. For instance, Fu et al. investigated the effects of LiF and Li_2CO_3 used as lithium sources on the properties of $Li_3V_2(PO_4)_3$. The results revealed that $Li_3V_2(PO_4)_3$ was prepared by using LiF as precursor at lower calcination temperature (700 $^\circ$C) and exhibited smaller particle size (about 300 nm), and thus displayed better electrochemical performance. The impact of different sintering temperatures on the performance of $Li_3V_2(PO_4)_3$ was also studied by Fu et al. and the results indicated that optimized temperature is 900 $^\circ$C and the optimal $Li_3V_2(PO_4)_3$ had the particle size of 1-5 μm and the highest discharge capacity. When calcination temperature is not higher than 750 °C, the synthesized product contained impurity of Li_3PO_4 but temperature is high as 1000 °C, the particles grow quite large (1-5 μm).

One drawback of the above-mentioned solid state reaction is to take a long period to calcination reactants by traditional heating method. To overcome this disadvantage, microwave heating, which can dramatically short reaction period from days and hours to minutes and seconds [11-14], has been introduced to synthesis of $Li_3V_2(PO_4)_3$ recently[15-17], and these $Li_3V_2(PO_4)_3$ samples show better electrochemical performance than those ones prepared by conventional traditional solid state reaction.

Although the above solid state reactions have merits of simple and suitable for large-scale production, they still suffer from a series of intractable issues, such as inhomogeneity, irregular morphology, uncontrollable particle growth and agglomeration, unfavorable for further improvement of electrochemical performance, especially at high rates, of $Li_3V_2(PO_4)_3$.

2.3.2 Sol-gel

Sol-gel synthesis is a wet chemical approach for producing solid materials from small, and is regarded as a promising route to prepare nanosized electroactive materials for lithium ion batteries [1] because the sol-gel method has a majority of advantages over conventional solid state reaction, such as homogeneous mixing reactants at the atomic or molecular level, good stoichiometric control, low synthesis temperature, short-heating time, good crystallinity, uniform particle size and small diameter, even down to nanometer scale [18]. The sol-gel process generally comprises the following steps [1]: precursor solution → hydrolysis → reactive monomer → condensation → sol gelation → gel → further treatment. On the basis of this procedure, all reactants are usually dissolved by a certain solvent or mixed solvents to form precursor solution, and ligands, different addition sequences of reactants and further treatment also influence the properties of the final products. In case of $Li_3V_2(PO_4)_3$, the reactants involve soluble lithium salts (e.g. LiOH, CH_3COOLi, and $LiNO_3$), phosphate salts (e.g. $NH_4H_2PO_4$), H_3PO_4 and ligands (e.g. citric acid, oxalic acid, poly(vinyl alcohol) (PVA), poly(vinylpyrrolidone) (PVP), and glycine). However, vanadium sources including V_2O_5 and NH_4VO_3 are difficult to be dissolved in distilled water or organic solvents and they were dissolved by the assistance of ligands. Taking citric acid as an example [19], citric acid is equivalent to V_2O_5 and stoichiometric amounts of $NH_4H_2PO_4$ and LiOH are added to the mixed solution of V_2O_5 and H_2O_2 under vigorous stirring, followed by evaporation of water at 60 °C to obtain a gel precursor. Then the precursor is calcined at 800 °C for 8 h under a stream of argon gas to get carbon coated $Li_3V_2(PO_4)_3$. Citric acid is used not only as the carbon source for reduction and coating but also as a chelating reagent. During the sol-gel process, citric acid can reduce V^{5+} to V^{4+} and V^{3+} as well reacts with the other reactants to form complexes. Upon calcination, citric acid can be pyrolyzed to carbon, and the carbon can reduce high valent vanadium to V^{3+}, coat the $Li_3V_2(PO_4)_3$ particles and suppress the growth of $Li_3V_2(PO_4)_3$ crystals. The as-prepared carbon coated $Li_3V_2(PO_4)_3$ presents a sphere-like particle morphology with size of about 400 nm and exhibits better electrochemical performance in the voltage range of 3.0-4.3 and 3.0-4.8 V (vs. Li^+/Li) at room temperature and elevated temperature (55 °C).

2.3.3 Hydrothermal Synthesis

Hydrothermal synthesis refers to the synthesis by chemical reactions of substance in a sealed solution above ambient temperature and pressure [20, 21]. Basically, the mechanism of hydrothermal reactions follows a liquid nucleation model, while the reaction mechanism of solid state reactions involves mainly diffusion of atoms or ions at the interface between reactants [21]. In hydrothermal synthesis, water, organic solvent or mixture of water and organic solvent are used as reaction medium or reactants. The pressure is commonly self-generated by solvents and the targeted reaction temperature is higher than the boiling point. The previous studies reveal that hydrothermal synthesis has the merits of homogeneous products, uniform particle size, morphology control, fast reaction kinetics, short reaction time, phase purity, high crystallinity, low temperature post-calcination, low cost, benign environment, etc., and has been widely applied to synthesis of nanostructural materials [1, 20-23].

According to the previous reports about hydrothermal synthesis of $Li_3V_2(PO_4)_3$/C, distilled water is used as reaction medium and the solution contains organic compounds. The targeted reaction temperatures are in the range of 160 to 180 °C. In order to improve the crystallinity and the electronic conductivity of coating carbon derived from pyrolysis of organic compounds, most of $Li_3V_2(PO_4)_3$ products prepared by hydrothermal reaction require post-calcination of high temperature. The flake-like $Li_3V_2(PO_4)_3$/C was prepared by hydrothermal reaction at 160 °C for 2 h and post-calcination of 700 °C for 6 h using NH_4VO_3, $NH_4H_2PO_4$, glucose and $LiOH \cdot H_2O$ as raw materials and exhibited initial discharge capacity of 164 mA/g at 5C in the voltage range of 3.0-4.8 V (vs. Li^+/Li) and better cycle performance, much better than $Li_3V_2(PO_4)_3$/C prepared by solid state reaction [24]. In 2010, $Li_3V_2(PO_4)_3$/C nanorods with diameters of about 60 nm and lengths of 0.5-1.0 μm prepared by hydrothermal reaction at 180 °C for 24-36 h without further calcination were reported by Liu et al., in which Li_2CO_3, $HCOOCOOH \cdot H_2O$, H_3PO_4 and V_2O_5 were used as starting materials. The nanorods displayed initial discharge capacity of 101.1 mAh/g at a current density of 1800 mA/g and excellent cycle performance in the voltage range of 3.0-4.6 V (vs. Li$^+$/Li) [25]. Sun et al. used NH$_4$VO$_3$, NH$_4$H$_2$PO$_4$, CH$_3$COOLi·2H$_2$O or Ca(C$_2$H$_3$O$_2$)$_2$ or (Sc(NO$_3$)$_3$·H$_2$O and ethylene glycol as raw materials to obtain Li$_3$V$_2$(PO$_4$)$_3$/C, Li$_{2.96}$Ca$_{0.02}$V$_2$(PO$_4$)$_3$/C and Li$_3$V$_{1.85}$Sc$_{0.15}$(PO$_4$)$_3$/C via hydrothermal reaction at 180 °C for 24 h and the followed calcination at 700 °C for 10 h, and these samples exhibited better electrochemical performance [26]. The plate-like Li$_3$V$_2$(PO$_4$)$_3$/C was prepared by calcination of glucose and Li$_3$V$_2$(PO$_4$)$_3$ precursor gel, which was obtained by hydrothermal reaction with NH$_4$VO$_3$, NH$_4$H$_2$PO$_4$, CH$_3$COOLi·2H$_2$O, HNO$_3$ and ethylene glycol as starting materials. The plate-like Li$_3$V$_2$(PO$_4$)$_3$/C delivers high capacity and excellent cyclability, even at low temperatures down to -20 °C, and it presents high capacity of 117.4 mAh/g and capacity retention rate of 97.2% after 80 cycles in the voltage range of 3.0-4.3 V (vs. Li$^+$/Li) [27]. Duan et al. prepared core-shelled structural Li$_3$V$_2$(PO$_4$)$_3$/C by hydrothermal method using V$_2$O$_5$, NH$_4$H$_2$PO$_4$, Li$_2$CO$_3$, ascorbic acid and polyethylene glycol (PEG-400) as raw materials, and the hydrothermal reaction was conducted at 180 °C for 40 h. The core-shelled Li$_3$V$_2$(PO$_4$)$_3$/C displays high capacity of 138 mAh/g at 5C and high capacity retention of 86% after 1000 cycles in the voltage range of 3.0-4.8 V (vs. Li$^+$/Li) [28]. In 2013, Pei et al. used mixture of ethanol and ethylene glycol as reaction medium and used V(C$_5$H$_7$O$_2$)$_3$, LiH$_2$PO$_4$ and reduced modified graphene oxide (rmGO) as reactants to prepare reduced graphene oxide supported Li$_3$V$_2$(PO$_4$)$_3$ by hydrothermal reaction at 180 °C for 24 h. The Li$_3$V$_2$(PO$_4$)$_3$/rmGO composite shows high capacity of 135 mAh/g and excellent cyclability at 5C in the voltage range of 3.0-4.8 V (vs. Li$^+$/Li).

2.3.4 Spray Pyrolysis

Spray pyrolysis is a simple and versatile technique for production of materials of a wide range of composition, size, and morphology [29], and it also has the advantages such as large scale production of high purity, homogeneous powders with fine size and no agglomeration. Basically, spray pyrolysis apparatus consists of a droplet generator, a quartz reactor and a powder collector [30]. In spray pyrolysis, precursor solution is firstly atomized by a droplet generator, and then the droplets are carried to a tabular heated reactor through carrier gas, where the solvent in the droplet is evaporated and solid particles are formed. Subsequently, the solid particles are calcined in the high temperature zone reaction tube to form the crystallized phases. The properties of particles synthesized by spray pyrolysis depend on the process parameters such as the composition of precursor solution, atomization of precursor solution, aerosol transport and decomposition of precursor.

Ko et al. reported that spherical carbon coated Li$_3$V$_2$(PO$_4$)$_3$ was prepared by ultrasonic spray pyrolysis method using Li$_2$CO$_3$, V$_2$O$_5$, NH$_4$H$_2$(PO$_4$)$_3$, sucrose and dilute HNO$_3$

solution as starting materials and the temperature of reactor was set to 1300 °C [31]. The precursor powders produced by spray pyrolysis exhibited spherical morphology and subsequently was sintered at 700 °C for 3 h in a mixture gas of 10% hydrogen and 90% nitrogen to obtain spherical coated $Li_3V_2(PO_4)_3$. The prepared sample presents capacity of 141 mAh/g and better cycle performance at 0.1C in the voltage range of 3.0-4.8 V (vs. Li^+/Li), much better than pristine $Li_3V_2(PO_4)_3$. After altering carbon sources and chelating reagent by addition of citric acid and ethylenediaminetetraacetic acid (EDTA), carbon coated hollow spherical $Li_3V_2(PO_4)_3$ was prepared by Ko et al. via spray pyrolysis and this sample displayed capacity of 147 mAh/g at better cycle performance at 0.1C in the voltage scope of 3.0-4.8 V (vs. Li^+/Li) [31].

2.3.5 Electrostatic Spray Deposition

Recently electrostatic spray deposition (ESD), also referred as electrostatic spray pyrolysis, has been regarded as a promising technique to manufacture inorganic thin films with various morphologies at nanoscale, such as dense, porous, cross-linked, and sponge-like structure, especially for electrode materials for lithium ion batteries [1, 32]. Compared with some conventional deposition techniques (*e.g.*, sputtering, chemical vapor deposition), ESD also has advantages such as a simple set-up, inexpensive, high deposition efficiency, and non-vacuum operation [33]. ESD is based on the spray formation from precursor solution and transportation of spray to substrate under the high direct electric field. A high direct current voltage is applied between an electrically conductive substrate such as metal or indium tin oxide (ITO) and a metal nozzle, which is connected to a precursor solution. When the direct electric field is increased to a certain onset the solution is atomized at orifice of the nozzle, hence, a spray is generated. Then the spray moves to the heated substrate under the electrostatic force and a thin layer is deposited on the substrate surface due to pyrolysis of the precursor [34]. The layer morphology is affected by many factors such as substrate temperature, flow speed, conductivity and viscosity of precursor solution and evaporation speed of solvent.

In 2010, Wang et al. reported that a $Li_3V_2(PO_4)_3$/C film was prepared by ESD technique using NH_4VO_3, $LiNO_3$, H_3PO_4, glucose, distilled water, ethanol, and 1,2-propylene glycol as raw materials [35]. The applied DC voltage was 12.5 kV and the distance between needle and substrate was 3 cm. The substrate was a flat graphite sheet and heated at 240 °C. The deposited film was finally calcined at 700 °C for 8 h in argon atmosphere to obtain $Li_3V_2(PO_4)_3$/C film, which consists of majority of small crystals (about 50 nm) anchored onto graphite sheet.

2.3.6 Electrospinning

Electrospinning is highly versatile method to turn solutions or melts, mainly of polymers or solutions containing polymers, into continuous fibers of various organic, inorganic and hybrid materials with diameters ranging from a few micrometers to a few nanometers [1]. The electrospinning has advantages of simple, versatile, low cost, high yield, and high degree of reproducibility of the prepared materials and it has been widely used to prepare nanofibrous materials [37, 38].The principle of electrospinning is based on the application of direct current electric field between metal nozzle of syringe and conductive collector. When the applied electric field strength increases to a critical value, the electrostatic repulsions among the charges on the surface of the drop are higher than the surface tension, and a jet is drawn from syringe under a constant flow rate, and the jet moves to the collector [6]. During the movement of the jet, jet is stretched into thinner jet and the solvent of jet evaporates, and hence the jet is solidified. Recently electrospinning technique has been employed to prepare nanofibrous electrode materials for lithium ion batteries [37-42]. In the case of $Li_3V_2(PO_4)_3$, a pioneering work about the electrospinning method was carried out by Chen et al. to prepare $Li_3V_2(PO_4)_3$/C nanofibers composite using NH_4VO_3, $NH_4H_2PO_4$, $CH_3COOLi \cdot 2H_2O$, citric acid and PVP (M_w=1,300,000) as raw materials [43]. The raw materials were dissolved by distilled water and form a clear viscous solution for electrospinning. After electrospinning process, the collected precursor nanofibers were calcined at 800 °C for 4 h to obtain $Li_3V_2(PO_4)_3$/C nanofibers composite with large surface area of 160.75 m^2/g and diameters of 90-220 nm. The nanofibers show high discharge capacity of 190 mAh/g at a current density of 19.7 mA/g (0.1 C) in the voltage range of 3.0-4.8 V (vs Li^+/Li) and excellent cycle performance. Even at a high rate of 20C, the nanofibers can deliver high discharge capacity of 132 mAh/g and better cyclability.

2.3.7 Other Methods

Some other approaches have also employed to prepare $Li_3V_2(PO_4)_3$/C composite except for the above-mentioned classic synthesis methods. One of them is freeze-drying method and the main advantage of this method is that homogeneity of precursor after removal of solvent from precursor solution under the vacuum and lower temperature. In 2012, Wang et al. prepare nanostructured $Li_3V_2(PO_4)_3$/C composite by freeze-drying method using NH_4VO_3, $LiOH \cdot H_2O$, H_3PO_4 and citric acid as raw materials [44]. The raw materials were dissolved by distilled water and formed a mixture of solution, and then the mixture of solution was dropped into liquid nitrogen, followed by vacuum drying process at -40 °C to remove the solid solvent. The freeze-drying precursor was calcined at high temperature to obtain nanostructured $Li_3V_2(PO_4)_3$/C composite, and this composite presented good high rate performance and cyclability between 3.0 and 4.3 V (vs Li^+/Li). Another similar freeze-drying work was reported by Qiao et al.[45]. The mixed solution of NH_4VO_3, $LiOH \cdot H_2O$ and H_3PO_4 was chilled to ice cake and then dried at -53 °C and vacuum to get rid of solvent. The resulted dried precursor was ground with polystyrene sphere, and then sintered at 750 °C for 8 h under argon flow to get nanostructured $Li_3V_2(PO_4)_3/C$ composite. This composite exhibited high discharge capacity of 93.3 mAh/g at 29.6 C in the voltage window of 3.0-4.3 V (vs Li^+/Li) and better cycle performance. Another synthesis method, rheological phase reaction (RPR), was firstly employed by Chang et al. to prepare $Li_3V_2(PO_4)_3/C$ composite [46]. Actually, PRP is a liquid-solid phase reaction in which solid reactants and liquid ones are dispersed homogeneously. In 2008, Chang et al. reported that $Li_3V_2(PO_4)_3/C$ composite was prepared by RPR method using Li_2CO_3, V_2O_5, $NH_4H_2PO_4$, polyethylene glycol (PEG) and distilled water as raw materials. A viscous rheological body was obtained by adjusting the amount of PEG and distilled water, and then was dried and calcined at high temperature under inert argon flow to obtain $Li_3V_2(PO_4)_3/C$ composite with better electrochemical performance than that of $Li_3V_2(PO_4)_3/C$ prepared by solid state reaction. Another RPR work was reported by Huang et al. using Li_2CO_3, NH_4VO_3, 1-Hydroxy Ethylidene-1,1-Diphosphonic acid (HEDP) as raw materials and the resulted $Li_3V_2(PO_4)_3/C$ composite exhibited better electrochemical performance [47]. A glass-ceramic technique, which is a simple, high-speed, and low cost synthesis method, was introduced by Nagamine et al. to prepare $Li_3V_2(PO_4)_3/C$ composite [48]. A glass with the composition of $37.5Li_2O-25V_2O_5-37.5P_2O_5$ (at molar ratio) was initially prepared by a conventional melt-quenching method and then ground to powder, subsequently mixed with glucose, calcined at 700 °C for 6 h under 7% H_2/Ar flow to obtain $Li_3V_2(PO_4)_3/C$ composite. An in situ polymerization method was also put forward by Mao et al. to synthesize $Li_3V_2(PO_4)_3/C$ composite using $(NH_4)_2S_2O_8$ as initiator, acrylamide as monomer, N,N'-methylene-bisacrylamide as crosslinking agent [1, 49]. During the formation of $Li_3V_2(PO_4)_3$ precursor sol, polymer reaction between monomer and crosslinking agent is occurred in situ, and the polymer is distributed in the mixture of precursor and polymer. After calcination of the mixture at high temperature, the polymer is pyrolyzed to 3D carbon network in situ, which is beneficial to significantly improve the electronic conductivity and hence to enhance the electrochemical performance of $Li_3V_2(PO_4)_3$.

2.4   Doping Elements in Lithium Vanadium Phosphate

The introduction of heteroatom into the crystal structure of electroactive materials will alter the energy band gap and the crystal cell, and hence will change the intrinsic electronic conductivity and ions diffusion in particles of electroactive materials. The positive effects of doping on the capacity, cycle life and rate capability of electroactive materials for lithium ion batteries have been intensively illustrated [50-53]. As for $Li_3V_2(PO_4)_3$, the doping is classified into the following three main types: (i) doping cations at V sites, (ii) doping cations at Li sites, (iii) doping anions at PO_4 sites.

2.4.1 Doping Cations at V Sites

The V in $Li_3V_2(PO_4)_3$ are partially substituted by divalent, trivalent or supervalent cations such as Fe^{3+}[54, 55], Al^{3+} [1], Cr^{3+} [56], Mg^{2+} [1, 57], Sc^{3+} [58], Co^{2+} [55, 59], Mn^{2+} [60], Ce^{3+}[61], Mo^{4+} [62], Sn^{4+} [63], Ni^{2+} [64], Mn^{3+} [65, 66], Zr^{4+} [67], Tm^{3+} [68], Ti^{4+} [69], and Zn^{2+} [70]. The $Li_3V_2(PO_4)_3$ is also codoped by Ti^{4+}-Mg^{2+}[71], Ti^{4+}-Mn^{2+} [72], and Ti^{4+}-Fe^{2+} [72]. The previous reports demonstrated that the above-mentioned dopants can increase the electrochemical performance of $Li_3V_2(PO_4)_3$.

The framework of $Li_3V_2(PO_4)_3$ is similar to those of $Li_3M_2(PO_4)_3$ (M=Ti, Fe, Cr, Al and Sc) compounds with different sizes of metal ions [1], which is favorable for substitution of vanadium by other metal ions. The doped $Li_3V_2(PO_4)_3$ was first reported in 2006 by Ren et al. using Fe^{3+} as a dopant and had the higher electronic conductivity and structural stability than pristine $Li_3V_2(PO_4)_3$ [54]. The optimal doped sample, $Li_3V_{1.98}Fe_{0.02}(PO_4)_3$, has high initial discharge capacity of 177 mAh/g at 0.2C and capacity retention of 71% after 80 cycles in the voltage scope of 3.0-4.9 V (vs. Li^+/Li), much better than initial capacity of 169 mAh/g and capacity retention of 58% for undoped $Li_3V_2(PO_4)_3$ (Fig.3a). While Nathiya et al. found that the optimal x for $Li_3V_{2-x}Fe_x(PO_4)_3$ was 0.05 in the x range of 0.05-0.15, and the optimal $Li_3V_{1.95}Fe_{0.05}(PO_4)_3$/C exhibited high capacity of 174 mAh/g at 0.1C and the capacity retention of 96%, much higher than that capacity retention of 86% for pristine $Li_3V_2(PO_4)_3$/C (Fig.3b) [55]. Except for Fe^{3+} dopant, the reported trivalent metal ions such as Al^{3+} [1], Cr^{3+} [56], Ce^{3+} [61], Mn^{3+} [66, 67] and Tm^{3+} [68] also can improve the electrochemical performance. Among these ions, Cr^{3+} was considered as a promising dopant and confirmed to increase the electronic conductivity of $Li_3V_2(PO_4)_3$. The electronic conductivity of $Li_3V_2(PO_4)_3$/C, $Li_3V_{1.95}Cr_{0.05}(PO_4)_3$/C, $Li_3V_{1.9}Cr_{0.1}(PO_4)_3$/C, $Li_3V_{1.8}Cr_{0.2}(PO_4)_3$/C and $Li_3V_{1.5}Cr_{0.5}(PO_4)_3$/C is $5.63×10^{-4}$, $2.17×10^{-3}$, $2.63×10^{-3}$, $3.45×10^{-3}$ and $4.33×10^{-3}$ S/cm, respectively. The optimal Cr-doped $Li_3V_2(PO_4)_3$/C, $Li_3V_{1.9}Cr_{0.1}(PO_4)_3$/C, exhibited high capacity of 171.4 mAh/g at 0.2C and better rate capability in the voltage range of 3.0-4.8 V (vs. Li^+/Li) (Fig.3c.) [56]. Among the above-mentioned reported tetravalent ions of Mo^{4+} [63], Sn^{4+} [64], Zr^{4+} [67], and Ti^{4+} [69], the Zr-doped $Li_3V_2(PO_4)_3$/C showed the best electrochemical performance and the optimal $Li_3V_{1.87}Zr_{0.1}(PO_4)_3$/C displayed initial discharge capacity of 181.2 mAh/g at 0.5C, better rate capability and cycle performance in the voltage range of 3.0-4.8 V (vs. Li^+/Li) (Fig.3d) [67]. Compared with the other above-mentioned bivalent dopants, the optimal Co^{2+} doped $Li_3V_{1.85}Co_{0.15}(PO_4)_3/C$ delivered initial discharge capacity of 163.3 mAh/g and better cycle performance in the voltage range of 3.0-4.8 V (vs. Li^+/Li) (Fig.3e) [59]. The combination of tetravalent ions and bivalent ions had also been applied in doped $Li_3V_2(PO_4)_3$ [71, 72]. The optimal Ti^{4+}-Mn^{2+} and Ti^{4+}-Fe^{2+} codoped $Li_3V_2(PO_4)_3$ showed high discharge capacity and better rate capability (Fig. 3f).

2.4.2 Doping Cations at Li Sites

The reported doping cations in Li sites of $Li_3V_2(PO_4)_3$ involve Ca^{2+} [73], Na^+ [1, 74] and Mg^{2+} [75]. The substitution of Li^+ by larger cations or high valent cations with similar radius can enlarge the cell volume of $Li_3V_2(PO_4)_3$ and provide more channel space for Li^+ transportation, resulting in higher Li^+ diffusion coefficient and better cycle performance. For instance, Na-doped $Li_{2.95}Na_{0.05}V_2(PO_4)_3/C$ composite prepared by a sol-gel method shows at least three times higher Li^+ diffusion coefficient than $Li_3V_2(PO_4)_3$ [76], which was verified by the calculation of galvanostatic intermittent titration technique (GITT) method. Moreover, the EIS results reveal that the charge transfer resistance is significantly reduced after Na^+-doping. As a result, the electrochemical performance of Na^+-doped $Li_3V_2(PO_4)_3/C$ composites is much better than that of $Li_3V_2(PO_4)_3/C$, and $Li_{2.95}Na_{0.05}V_2(PO_4)_3/C$ composite with 2.9 wt% carbon shows the highest discharge capacity and the best cyclability. The $Li_{2.95}Na_{0.05}V_2(PO_4)_3/C$ composite presents an initial capacity of 173.1 mAh/g and capacity retention of 91% after 30 cycles at a high rate of 1C in the voltage range of 3.0-4.8 V (vs. Li^+/Li) (Fig. 3g).

2.4.3 Doping Anions at PO_4 Sites

In addition to the above-mentioned doped cations, anions have also been investigated as dopants to improve the electrochemical performance of $Li_3V_2(PO_4)_3/C$. The substitution of F^- for PO_4^{3-} was reported to improve the electronic conductivity and enhance the electrochemical performance of $Li_3V_2(PO_4)_3/C$. The optimal F^- doped $Li_3V_2(PO_4)_3/C$ sample, $Li_3V_2(PO_4)_{2.9}F_{0.1}$ displays high electronic conductivity of 7.2×10^{-6} S/cm and initial discharge capacity of 117 mAh/g at 10C in the voltage range of 3.0-4.3 V, while the electronic conductivity and initial discharge capacity at 10 C of $Li_3V_2(PO_4)_3/C$ are only 3.7×10^{-8} S/cm and 74 mAh/g, respectively (Fig.3h) [77]. Another anion Cl^- was also introduced into $Li_3V_2(PO_4)_3$ and the doped $Li_3V_2(PO_4)_{3-x}Cl_x$ showed much higher Li^+ diffusion coefficients than the pristine $Li_3V_2(PO_4)_3$, which may result from that fact that the strong electronegative of Cl^- could decrease the Li-O bond energy and result in facile extraction of Li^+ from the host [78].

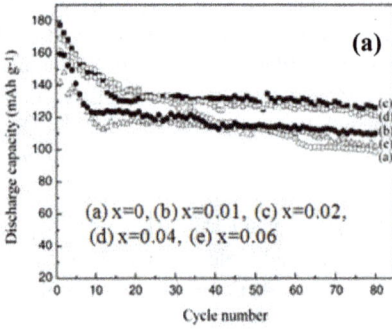

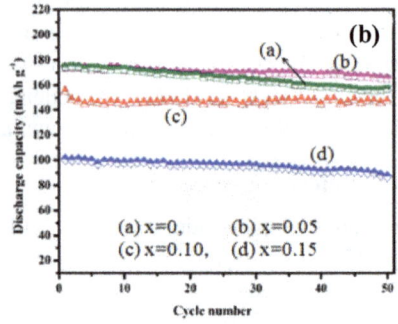

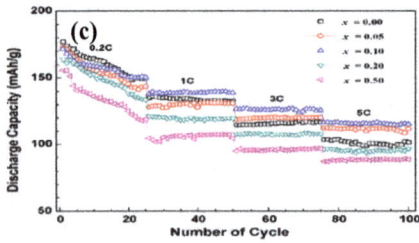

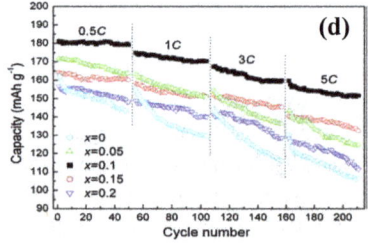

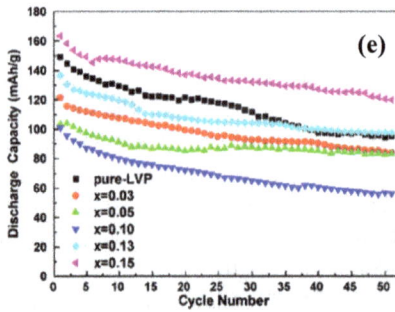

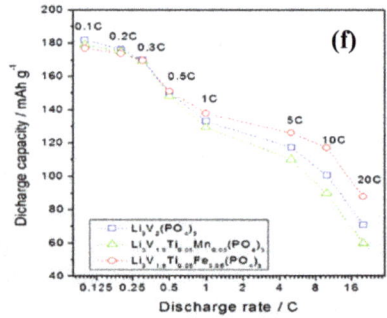

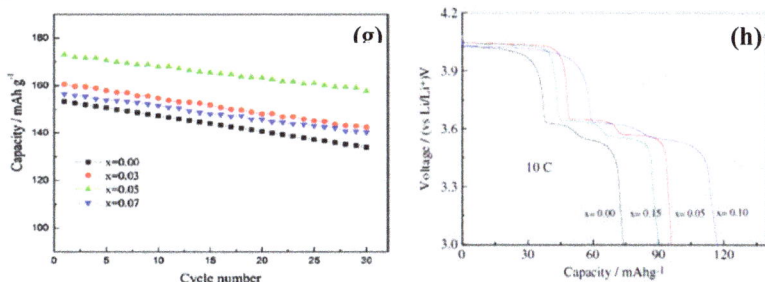

Figure 3. (a) Cycle performance of $Li_3V_{2-x}Fe_x(PO_4)_3/C$ at 0.2 C. Reproduced with permission from Ref. [54]. (b) Cycle performance of $Li_3V_{2-x}Fe_x(PO_4)_3/C$ at 0.1 C. Reproduced with permission from Ref. [55]. (c) Rate capability and cycle performance of $Li_3V_{2-x}Cr_x(PO_4)_3/C$. Reproduced with permission from Ref.[56]. (d) Rate capability and cycle performance of $Li_3V_{2-x}Zr_x(PO_4)_3/C$. Reproduced with permission from Ref.[67]. (e) Cycle performance of $Li_3V_{2-x}Co_x(PO_4)_3/C$. Reproduced with permission from Ref.[59]. (f) Rate capability of $Li_3V_{2-2x}Ti_xMn_x(PO_4)_3/C$ and $Li_3V_{2-2x}Ti_xFe_x(PO_4)_3/C$. Reproduced with permission from Ref.[71, 72]. (g) Cycle performance of $Li_{3-x}Na_xV_2(PO_4)_3$ at 1C. Reproduced with permission from Ref. [76]. (h) Initial capacity of $Li_3V_2(PO_4)_{3-x}F_x$ at 10C. Reproduced with permission from Ref.[77].

2.5 Surface Modifications

Surface modification technique has been widely applied to electrochemical performance of electrode materials for lithium ion batteries. The surface modifications of the electroactive material particles generally include the coating of the particles with a layer that protects the core region from side reactions with electrolyte, and prevents the loss of oxygen, and the dissolution of the metal ions in the electrolyte, or simply improves the conductivity of the powder [79]. In the case of $Li_3V_2(PO_4)_3$, the surface modification is mainly to improve the electronic conductivity of $Li_3V_2(PO_4)_3$ by coating of conductive materials such as carbon and silver [80]. The carbon coating is the most common way to enhance the electronic conductivity of $Li_3V_2(PO_4)_3$ and hence to improve the electrochemical performance. The carbon coating is usually achieved by introduction of carbon sources in the starting materials, and the carbon sources always include carbon materials such as carbon black, high area carbon, graphene, carbon nanoflakes, etc., and organic precursors such as citric acid, ascorbic acid, humic acid, and so on. Compared with the carbon materials, organic precursors are more effective because organic precursors easily form the dispersive carbon in situ during the process of high temperature calcination and the dispersive carbon is prone to suppress the growth of

$Li_3V_2(PO_4)_3$ particles, resulting in $Li_3V_2(PO_4)_3/C$ composite with smaller particles and high electronic conductivity. As a result, the carbon coated $Li_3V_2(PO_4)_3$ prepared from the starting materials containing organic precursors exhibit better electrochemical performance.

2.6　Summary

In recent years, monoclinic $Li_3V_2(PO_4)_3$ possessing excellent cyclability, high theoretical capacity of 197 mAh/g, low synthetic cost, excellent thermal stability and environmental friendliness has become a research focus of cathode materials for lithium ion batteries, and has been considered as a highly suitable candidate for the cathodes of the next-generation lithium ion batteries.

References

[1]　S.C. Yin, H. Grondey, P. Strobel, M. Anne, L.F. Nazar. Electrochemical Property: Structure Relationships in Monoclinic Li3-yV2(PO4)3, Journal of the American Chemical Society, 2003,125, 10402-10411. https://doi.org/10.1021/ja034565h

[2]　M.Y. Saïdi, J. Barker, H. Huang, J.L. Swoyer, G. Adamson. Electrochemical Properties of Lithium Vanadium Phosphate as a Cathode Material for Lithium-Ion Batteries. Electrochemical and Solid-State Letters,2002, 5, A149-A151. https://doi.org/10.1149/1.1479295

[3]　J. Gaubicher, C. Wurm, G. Goward, C. Masquelier, L. Nazar. Rhombohedral Form of Li3V2(PO4)3 as a Cathode in Li-Ion Batteries. Chemistry of Materials, 2000, 12, 3240-3242. https://doi.org/10.1021/cm000345g

[4]　H. Huang, S.C. Yin, T. Kerr, N. Taylor, L.F. Nazar, Nanostructured Composites: A High Capacity, Fast Rate Li3V2(PO4)3/Carbon Cathode for Rechargeable Lithium Batteries. Advanced Materials, 2002, 14, 1525-1528. https://doi.org/10.1002/1521-4095(20021104)14:21<1525::AID-ADMA1525>3.0.CO;2-3

[5]　M. Sato, H. Ohkawa, K. Yoshida, M. Saito, K. Uematsu, K. Toda. Enhancement of discharge capacity of Li3V2(PO4)3 by stabilizing the orthorhombic phase at room temperature. Solid State Ionics, 2000, 135, 137-142. https://doi.org/10.1016/S0167-2738(00)00292-7

[6]　X. Rui, Q. Yan, M. Skyllas-Kazacos, T.M. Lim. Li3V2(PO4)3 cathode materials for lithium-ion batteries: A review. Journal of Power Sources, 2014, 258, 19-38. https://doi.org/10.1016/j.jpowsour.2014.01.126

[7]   M.Y. Saïdi, J. Barker, H. Huang, J.L. Swoyer, G. Adamson. Performance characteristics of lithium vanadium phosphate as a cathode material for lithium-ion batteries. Journal of Power Sources, 2003, 119-121, 266-272. https://doi.org/10.1016/S0378-7753(03)00245-3

[8]   H. Liu, G. Yang, X. Zhang, P. Gao, L. Wang, J. Fang, J. Pinto, X. Jiang. Kinetics of conventional carbon coated-Li3V2(PO4)3 and nanocomposite Li3V2(PO4)3/graphene as cathode materials for lithium ion batteries. Journal of Materials Chemistry, 2012, 22, 11039-11047. https://doi.org/10.1039/c2jm31004j

[9]   H. Tukamoto, A.R. West. Electronic Conductivity of LiCoO2 and Its Enhancement by Magnesium Doping. Journal of The Electrochemical Society, 1997, 144, 3164-3168. https://doi.org/10.1149/1.1837976

[10]   M. Nishizawa, T. Ise, H. Koshika, T. Itoh, I. Uchida. Electrochemical In-Situ Conductivity Measurements for Thin Film of Li1-xMn2O4 Spinel. Chemistry of Materials, 2000, 12, 1367-1371. https://doi.org/10.1021/cm990696z

[11]   S. Beninati, L. Damen, M. Mastragostino. MW-assisted synthesis of LiFePO4 for high power applications. Journal of Power Sources, 2008, 180, 875-879. https://doi.org/10.1016/j.jpowsour.2008.02.066

[12]   G. Yang, G. Wang, W. Hou. Microwave solid-state synthesis of LiV3O8 as cathode material for lithium batteries. The Journal of Physical Chemistry B, 2005, 109, 11186-11196. https://doi.org/10.1021/jp050448s

[13]   Y. Qiao, X. Hu, Y. Liu, Y. Huang. Li4Ti5O12 nanocrystallites for high-rate lithium-ion batteries synthesized by a rapid microwave-assisted solid-state process. Electrochimica Acta, 2012, 63, 118-123. https://doi.org/10.1016/j.electacta.2011.12.064

[14]   K.J. Rao, B. Vaidhyanathan, M. Ganguli, P.A. Ramakrishnan. Synthesis of inorganic solids using microwaves. Chemistry of Materials,1999, 11, 882-895. https://doi.org/10.1021/cm9803859

[15]   G. Yang, H. Liu, H. Ji, Z. Chen, X. Jiang. Microwave solid-state synthesis and electrochemical properties of carbon-free Li3V2(PO4)3 as cathode materials for lithium batteries. Electrochimica Acta, 2010, 55, 2951-2957. https://doi.org/10.1016/j.electacta.2009.11.102

[16]   G. Yang, H. Liu, H. Ji, Z. Chen, X. Jiang. Temperature-controlled microwave solid-state synthesis of Li3V2(PO4)3 as cathode materials for lithium batteries.

Journal of Power Sources, 2010, 195, 5374-5378.
https://doi.org/10.1016/j.jpowsour.2010.03.037

[17]   G. Yang, H. Ji, H. Liu, B. Qian, X. Jiang. Crystal structure and electrochemical
       performance of Li3V2(PO4)3 synthesized by optimized microwave solid-state
       synthesis route. Electrochimica Acta, 2010, 55, 3669-3680.
       https://doi.org/10.1016/j.electacta.2010.01.114

[18]   L.J. Fu, H. Liu, C. Li, Y.P. Wu, E. Rahm, R. Holze, H.Q. Wu. Electrode materials
       for lithium secondary batteries prepared by sol–gel methods. Progress in Materials
       Science, 2005, 50, 881-928. https://doi.org/10.1016/j.pmatsci.2005.04.002

[19]   Q. Chen, J. Wang, Z. Tang, W. He, H. Shao, J. Zhang, Electrochemical
       performance of the carbon coated Li3V2(PO4)3 cathode material synthesized by a
       sol–gel method. Electrochimica Acta,2007, 52, 5251-5257.
       https://doi.org/10.1016/j.electacta.2007.02.039

[20]   K. Tekin, S. Karagöz, S. Bektaş. A review of hydrothermal biomass processing.
       Renewable and Sustainable Energy Reviews, 2014, 40, 673-687.
       https://doi.org/10.1016/j.rser.2014.07.216

[21]   S. Feng, R. Xu. New Materials in Hydrothermal Synthesis. Accounts of Chemical
       Research, 2001, 34, 239-247. https://doi.org/10.1021/ar0000105

[22]   W. Shi, S. Song, H. Zhang. Hydrothermal synthetic strategies of inorganic
       semiconducting nanostructures. Chemical Society Reviews, 2013, 42, 5714-5743.
       https://doi.org/10.1039/c3cs60012b

[23]   G. Demazeau, A. Largeteau. Hydrothermal/Solvothermal Crystal Growth: an Old
       but Adaptable Process. zeitschrift fur anorganische und allgemeine chemie, 2015,
       641, 159-163.

[24]   C. Chang, J. Xiang, X. Shi, X. Han, L. Yuan, J. Sun. Hydrothermal synthesis of
       carbon-coated lithium vanadium phosphate. Electrochimica Acta, 2008, 54, 623-
       627. https://doi.org/10.1016/j.electacta.2008.07.038

[25]   H. Liu, C. Cheng, X. Huang, J. Li. Hydrothermal synthesis and rate capacity
       studies of Li3V2(PO4)3 nanorods as cathode material for lithium-ion batteries.
       Electrochimica Acta, 2010, 55, 8461-8465.
       https://doi.org/10.1016/j.electacta.2010.07.049

[26]   C. Sun, S. Rajasekhara, Y. Dong, J.B. Goodenough. Hydrothermal synthesis and
       electrochemical properties of Li3V2(PO4)3/C-based composites for lithium-ion

batteries. ACS Applied Materials and Interfaces, 2011, 3, 3772-3776. https://doi.org/10.1021/am200987y

[27]   F. Teng, Z.-H. Hu, X.-H. Ma, L.-C. Zhang, C.-X. Ding, Y. Yu, C.-H. Chen. Hydrothermal synthesis of plate-like carbon-coated Li3V2(PO4)3 and its low temperature performance for high power lithium ion batteries. Electrochimica Acta, 2013, 91, 43-49. https://doi.org/10.1016/j.electacta.2012.12.090

[28]   W. Duan, Z. Hu, K. Zhang, F. Cheng, Z. Tao, J. Chen. Li3V2(PO4)3@C core-shell nanocomposite as a superior cathode material for lithium-ion batteries. Nanoscale, 2013, 5, 6485-6490. https://doi.org/10.1039/c3nr01617j

[29]   G.L. Messing, S.-C. Zhang, G.V. Jayanthi. Ceramic Powder Synthesis by Spray Pyrolysis. Journal of the American Ceramic Society,1993, 76, 2707-2726. https://doi.org/10.1111/j.1151-2916.1993.tb04007.x

[30]   D.S. Jung, Y.N. Ko, Y.C. Kang, S.B. Park. Recent progress in electrode materials produced by spray pyrolysis for next-generation lithium ion batteries. Advanced Powder Technology, 2014, 25, 18-31. https://doi.org/10.1016/j.apt.2014.01.012

[31]   Y.N. Ko, H.Y. Koo, J.H. Kim, J.H. Yi, Y.C. Kang, J.H. Lee. Characteristics of Li3V2(PO4)3/C powders prepared by ultrasonic spray pyrolysis. Journal of Power Sources,2011, 196, 6682-6687. https://doi.org/10.1016/j.jpowsour.2010.11.086

[32]   Y.N. Ko, J.H. Kim, Y.J. Hong, Y.C. Kang. Electrochemical properties of nano-sized Li3V2(PO4)3/C composite powders prepared by spray pyrolysis from spray solution with chelating agent. Materials Chemistry and Physics,2011, 131, 292-296. https://doi.org/10.1016/j.matchemphys.2011.09.044

[33]   C. Chen, E.M. Kelder, P.J.J.M. van der Put, J. Schoonman. Morphology control of thin LiCoO2 films fabricated using the electrostatic spray deposition (ESD) technique. Journal of Materials Chemistry,1996, 6, 765-771. https://doi.org/10.1039/jm9960600765

[34]   C.H. Chen, E.M. Kelder, M.J.G. Jak, J. Schoonman. Electrostatic spray deposition of thin layers of cathode materials for lithium battery. Solid State Ionics,1996, 86-88, Part 2, 1301-1306. https://doi.org/10.1016/0167-2738(96)00305-0

[35]   C.H. Chen, A.A.J. Buysman, E.M. Kelder, J. Schoonman. Fabrication of LiCoO2 thin film cathodes for rechargeable lithium battery by electrostatic spray pyrolysis. Solid State Ionics, 1995, 80, 1-4. https://doi.org/10.1016/0167-2738(95)00140-2

[36]   L. Wang, L.-C. Zhang, I. Lieberwirth, H.-W. Xu, C.-H. Chen. A Li3V2(PO4)3/C thin film with high rate capability as a cathode material for lithium-ion batteries.

Electrochemistry communications, 2010, 12, 52-55.
https://doi.org/10.1016/j.elecom.2009.10.034

[37] S. Cavaliere, S. Subianto, I. Savych, D.J. Jones, J. Roziere. Electrospinning: designed architectures for energy conversion and storage devices. Energy & Environmental Science, 2011, 4, 4761-4785. https://doi.org/10.1039/c1ee02201f

[38] Z. Dong, S.J. Kennedy, Y. Wu. Electrospinning materials for energy-related applications and devices. Journal of Power Sources, 2011, 196, 4886-4904. https://doi.org/10.1016/j.jpowsour.2011.01.090

[39] O. Toprakci, L. Ji, Z. Lin, H.A.K. Toprakci, X. Zhang. Fabrication and electrochemical characteristics of electrospun LiFePO4/carbon composite fibers for lithium-ion batteries. Journal of Power Sources, 2011, 196, 7692-7699. https://doi.org/10.1016/j.jpowsour.2011.04.031

[40] Q. Chen, X. Qiao, C. Peng, T. Zhang, Y. Wang, X. Wang. Electrochemical performance of electrospun LiFePO4/C submicrofibers composite cathode material for lithium ion batteries. Electrochimica Acta, 2012, 78, 40-48. https://doi.org/10.1016/j.electacta.2012.05.143

[41] M. Inagaki, Y. Yang, F. Kang. Carbon Nanofibers Prepared via Electrospinning. Advanced Materials, 2012, 24, 2547-2566. https://doi.org/10.1002/adma.201104940

[42] L. Ji, Z. Lin, M. Alcoutlabi, O. Toprakci, Y. Yao, G. Xu, S. Li, X. Zhang. Electrospun carbon nanofibers decorated with various amounts of electrochemically-inert nickel nanoparticles for use as high-performance energy storage materials. RSC Advances, 2012, 2, 192-198. https://doi.org/10.1039/C1RA00676B

[43] Q. Chen, T. Zhang, X. Qiao, D. Li, J. Yang. Li3V2(PO4)3/C nanofibers composite as a high performance cathode material for lithium-ion battery. Journal of Power Sources, 2013, 234, 197-200. https://doi.org/10.1016/j.jpowsour.2013.01.164

[44] C. Wang, H. Liu, W. Yang. An integrated core-shell structured Li3V2(PO4)3@C cathode material of LIBs prepared by a momentary freeze-drying method. Journal of Materials Chemistry, 2012, 22, 5281-5285. https://doi.org/10.1039/c2jm16417e

[45] Y.Q. Qiao, X.L. Wang, Y.J. Mai, X.H. Xia, J. Zhang, C.D. Gu, J.P. Tu. Freeze-drying synthesis of Li3V2(PO4)3/C cathode material for lithium-ion batteries. Journal of Alloys and Compounds, 2012, 536, 132-137. https://doi.org/10.1016/j.jallcom.2012.04.118

[46]   C. Chang, J. Xiang, X. Shi, X. Han, L. Yuan, J. Sun. Rheological phase reaction synthesis and electrochemical performance of Li3V2(PO4)3/carbon cathode for lithium ion batteries. Electrochimica Acta, 2008, 53, 2232-2237. https://doi.org/10.1016/j.electacta.2007.09.038

[47]   J.S. Huang, L. Yang, K.Y. Liu. Organic phosphoric sources for syntheses of Li3V2(PO4)3/C via improved rheological phase reaction. Materials Letters, 2012, 66,196-198. https://doi.org/10.1016/j.matlet.2011.08.097

[48]   K. Nagamine, T. Honma, T. Komatsu, A fast synthesis of Li3V2(PO4)3 crystals via glass-ceramic processing and their battery performance. Journal of Power Sources, 2011, 196, 9618-9624. https://doi.org/10.1016/j.jpowsour.2011.06.094

[49]   W.-f. Mao, J. Yan, H. Xie, Z.-y. Tang, Q. Xu. The interval high rate discharge behavior of Li3V2(PO4)3/C cathode based on in situ polymerization method. Electrochimica Acta, 2013, 88, 429-435. https://doi.org/10.1016/j.electacta.2012.10.078

[50]   T.F. Yi, Y. Xie, M.F. Ye, L.J. Jiang, R.S. Zhu, Y.R. Zhu. Recent developments in the doping of LiNi0.5Mn1.5O4 cathode material for 5 V lithium-ion batteries. Ionics, 2011, 17, 383-389. https://doi.org/10.1007/s11581-011-0550-6

[51]   T.F. Yi, X.Y. Li, H. Liu, J. Shu, Y.R. Zhu, R.S. Zhu. Recent developments in the doping and surface modification of LiFePO4 as cathode material for power lithium ion battery. Ionics, 2012, 18, 529-539. https://doi.org/10.1007/s11581-012-0695-y

[52]   A. Bhaskar, D. Mikhailova, N. Kiziltas-Yavuz, K. Nikolowski, S. Oswald, N.N. Bramnik, H. Ehrenberg. 3d-Transition metal doped spinels as high-voltage cathode materials for rechargeable lithium-ion batteries. Progress in Solid State Chemistry, 2014, 42, 128-148. https://doi.org/10.1016/j.progsolidstchem.2014.04.007

[53]   Q. Zhang, X. Li. Recent developments in the doped-Li4Ti5O12 anode materials of lithium-ion batteries for improving the rate capability. International Journal of Electrochemical Science, 2013, 8, 6449-6456.

[54]   M. Ren, Z. Zhou, Y. Li, X.P. Gao, J. Yan. Preparation and electrochemical studies of Fe-doped Li3V2(PO4)3 cathode materials for lithium-ion batteries. Journal of Power Sources, 2006, 162, 1357-1362. https://doi.org/10.1016/j.jpowsour.2006.08.008

[55]   K. Nathiya, D. Bhuvaneswari, Gangulibabu, D. Nirmala, N. Kalaiselvi. Li3MxV2-x(PO4)3/C (M=Fe, Co) composite cathodes with extended solubility limit and

improved electrochemical behavior. RSC Advances, 2012, 2, 6885-6889. https://doi.org/10.1039/c2ra20998e

[56]   Y. Chen, Y. Zhao, X. An, J. Liu, Y. Dong, L. Chen. Preparation and electrochemical performance studies on Cr-doped Li3V2(PO4)3 as cathode materials for lithium-ion batteries. Electrochimica Acta, 2009, 54, 5844-5850. https://doi.org/10.1016/j.electacta.2009.05.041

[57]   C. Dai, Z. Chen, H. Jin, X. Hu. Synthesis and performance of Li3(V1-xMgx)2(PO4)3 cathode materials. Journal of Power Sources, 2010, 195, 5775-5779. https://doi.org/10.1016/j.jpowsour.2010.02.081

[58]   Y.G. Mateyshina, N.F. Uvarov. Electrochemical behavior of Li3-xM′xV2-yM″y(PO4)3 (M′ =K, M″=Sc, Mg+Ti)/C composite cathode material for lithium-ion batteries. Journal of Power Sources, 2011, 196, 1494-1497. https://doi.org/10.1016/j.jpowsour.2010.08.078

[59]   Q. Kuang, Y. Zhao, X. An, J. Liu, Y. Dong, L. Chen. Synthesis and electrochemical properties of Co-doped Li3V2(PO4)3 cathode materials for lithium-ion batteries. Electrochimica Acta, 2010, 55, 1575-1581. https://doi.org/10.1016/j.electacta.2009.10.028

[60]   M. Bini, S. Ferrari, D. Capsoni, V. Massarotti. Mn influence on the electrochemical behaviour of Li3V2(PO4)3 cathode material. Electrochimica Acta, 2011, 56, 2648-2655. https://doi.org/10.1016/j.electacta.2010.12.011

[61]   J. Yao, S. Wei, P. Zhang, C. Shen, K.-F. Aguey-Zinsou, L. Wang. Synthesis and properties of Li3V2-xCex(PO4)3/C cathode materials for Li-ion batteries. Journal of Alloys and Compounds, 2012, 532, 49-54. https://doi.org/10.1016/j.jallcom.2012.04.014

[62]   W. Yuan, J. Yan, Z. Tang, O. Sha, J. Wang, W. Mao, L. Ma. Mo-doped Li3V2(PO4)3/C cathode material with high rate capability and long term cyclic stability. Electrochimica Acta, 2012, 72, 138-142. https://doi.org/10.1016/j.electacta.2012.04.030

[63]   G. Bai, Y. Yang, H. Shao, Synthesis and electrochemical properties of polyhedron-shaped Li3V2-xSnx(PO4)3 as cathode material for lithium-ion batteries. Journal of Electroanalytical Chemistry, 2013, 688, 98-102. https://doi.org/10.1016/j.jelechem.2012.08.018

[64]   W.L. Wu, J. Liang, J. Yan, W.F. Mao. Synthesis of Li3NixV2-x(PO4)3/C cathode materials and their electrochemical performance for lithium ion batteries. Journal

of Solid State Electrochem, 2013, 17, 2027-2033. https://doi.org/10.1007/s10008-013-2049-8

[65]   L. Chen, C. Wang, H. Wang, E. Qiao, S. Wang, X. Jiang, G. Yang. Enhanced high-rate electrochemical performance of Li3V1.8Mn0.2(PO4)3 by atomic doping of Mn(III). Electrochimica Acta, 2014, 125, 338-346. https://doi.org/10.1016/j.electacta.2014.01.118

[66]   L. Chen, B. Yan, Y. Xie, S. Wang, X. Jiang, G. Yang. Preparation and electrochemical properties of Li3V1.8Mn0.2(PO4)3 doped via different Mn sources. Journal of Power Sources, 2014, 261, 188-197. https://doi.org/10.1016/j.jpowsour.2014.03.061

[67]   J. Xu, G. Chen, H. Zhang, W. Zheng, Y. Li. Electrochemical performance of Zr-doped Li3V2(PO4)3/C composite cathode materials for lithium ion batteries. Journal of Applied Electrochemistry, 2014, 45, 123-130. https://doi.org/10.1007/s10800-014-0782-z

[68]   X. Yang, L. Jun, H. Jia, Study on structure and electrochemical performance of Tm3+-doped monoclinic Li3V2(PO4)3/C cathode material for lithium-ion batteries. Electrochimica Acta, 2014, 150, 62-67. https://doi.org/10.1016/j.electacta.2014.10.133

[69]   S.M. Stankov, I. Abrahams, A. Momchilov, I. Popov, T. Stankulov, A. Trifonova. Effect of Ti-doping on the electrochemical performance of lithium vanadium(III) phosphate. Ionics, 2015, 21, 1501-1508. https://doi.org/10.1007/s11581-014-1325-7

[70]   W. Wang, J. Zhang, Y. Lin, F. Ding, Z. Chen, C. Dai. A new carbon additive compounded Li3V1.97Zn0.05(PO4)3/C cathode for plug-in hybrid electric vehicles. Electrochimica Acta, 2015, 170, 269-275. https://doi.org/10.1016/j.electacta.2015.04.163

[71]   C. Deng, S. Zhang, S.Y. Yang, Y. Gao, B. Wu, L. Ma, B.L. Fu, Q. Wu, F.L. Liu. Effects of Ti and Mg Codoping on the Electrochemical Performance of Li3V2(PO4)3 Cathode Material for Lithium Ion Batteries. The Journal of Physical Chemistry C, 2011, 115, 15048-15056. https://doi.org/10.1021/jp201686g

[72]   S. Zhang, Q. Wu, C. Deng, F.L. Liu, M. Zhang, F.L. Meng, H. Gao, Synthesis and characterization of Ti-Mn and Ti-Fe codoped Li3V2(PO4)3 as cathode material for lithium ion batteries. Journal of Power Sources, 2012, 218, 56-64. https://doi.org/10.1016/j.jpowsour.2012.06.002

[73] C. Sun, S. Rajasekhara, Y. Dong, J.B. Goodenough. Hydrothermal Synthesis and Electrochemical Properties of Li3V2(PO4)3/C-Based Composites for Lithium-Ion Batteries. ACS Applied Materials & Interfaces, 2011, 3, 3772-3776. https://doi.org/10.1021/am200987y

[74] Q. Kuang, Y. Zhao, Z. Liang, Synthesis and electrochemical properties of Na-doped Li3V2(PO4)3 cathode materials for Li-ion batteries. Journal of Power Sources, 2011, 196, 10169-10175. https://doi.org/10.1016/j.jpowsour.2011.08.044

[75] W. Yin, T. Zhang, Q. Chen, G. Li, L. Zhang. Synthesis and electrochemical performance of Li3-2xMgxV2(PO4)3/C composite cathode materials for lithium-ion batteries. Transactions of Nonferrous Metals Society of China,2015, 25,1978-1985. https://doi.org/10.1016/S1003-6326(15)63806-7

[76] Q. Chen, X. Qiao, Y. Wang, T. Zhang, C. Peng, W. Yin, L. Liu. Electrochemical performance of Li3−xNaxV2(PO4)3/C composite cathode materials for lithium ion batteries. Journal of Power Sources, 2012, 201, 267-273. https://doi.org/10.1016/j.jpowsour.2011.10.133

[77] S. Zhong, L. Liu, J. Liu, J. Wang, J. Yang, High-rate characteristic of F-substitution cathode materials for Li–ion batteries. Solid State Communications, 2009, 149, 1679-1683. https://doi.org/10.1016/j.ssc.2009.06.019

[78] J. Yan, W. Yuan, Z.Y. Tang, H. Xie, W.F. Mao, L. Ma. Synthesis and electrochemical performance of Li3V2(PO4)3-xClx/C cathode materials for lithium-ion batteries. Journal of Power Sources, 2012, 209, 251-256. https://doi.org/10.1016/j.jpowsour.2012.02.110

[79] A. Mauger, C. Julien. Surface modifications of electrode materials for lithium-ion batteries: Status and trends. Ionics, 2014, 20, 751-787. https://doi.org/10.1007/s11581-014-1131-2

[80] L. Zhang, X.L. Wang, J.Y. Xiang, Y. Zhou, S.J. Shi, J.P. Tu. Synthesis and electrochemical performances of Li3V2(PO4)3/(Ag+C) composite cathode. Journal of Power Sources, 2010, 195, 5057-5061. https://doi.org/10.1016/j.jpowsour.2010.02.014

CHAPTER 4

Improvements of $Li_4Ti_5O_{12}$ Anode Material for Lithium-Ion Batteries

Chunfu Lin

State Key Laboratory of Marine Resource Utilization in South China Sea, Key Laboratory of Ministry of Education for Advanced Materials in Tropical Island Resources, College of Materials and Chemical Engineering, Hainan University, Haikou 570228, Hainan, PR China

linchunfu@hainu.edu.cn

Abstract

$Li_4Ti_5O_{12}$ is a promising anode material for lithium-ion batteries due to its good safety performance and excellent cyclic stability. However, it suffers from three problems of poor (electronic and ionic) conductivity, small capacity and gassing. This chapter reviews the characteristics, the crystal structure, the electrochemical performances and the improvements of $Li_4Ti_5O_{12}$. An insight into future research directions for $Li_4Ti_5O_{12}$ is also provided.

Keywords

Lithium-Ion Battery, Anode, $Li_4Ti_5O_{12}$, LIB Improvement, Electrochemical Performance

Contents

1.   Introduction

By virtue of high energy density, low self-discharge, long life span and absence of memory effect, lithium-ion batteries (LIBs) are not only popular in consumer electronics but also in electric vehicles (EVs) [1]. Graphite is the most popular anode material in consumer electronics due to its large capacity and low cost. It has a large theoretical capacity of 372 mAh g^{-1}. In general, it can deliver a practical capacity of >300 mAh g^{-1}. However, graphite may not be a desirable intercalation-type anode material in EVs due to its following disadvantages [2]. Its extremely low working potential of ~0.1 V causes the formation of lithium dendrites and thick solid-electrolyte interface (SEI) layers on its surface. Moreover, its Li^+-ion diffusion coefficient is limited. When the operation current density is large, it suffers from severe polarization causing the growth of lithium dendrites. Therefore, graphite cannot meet the requirements of good safety performance and high power density for EVs.

To develop the LIBs for EVs, it is highly desirable to explore new anode materials. So far, many new anode material, such as intercalation-type TiO_2 and $Li_4Ti_5O_{12}$, conversion-type CoO and Fe_2O_3, and alloying-type Sn and Si, have been explored [3]. Among them, $Li_4Ti_5O_{12}$ may have received the most attention since it has the following unique merits. $Li_4Ti_5O_{12}$ has a relatively high and flat working potential of ~1.55 V. Such high working potential avoids the formation of SEI layers and lithium dendrites, which can respectively benefit the rate performance and address safety issues. The flat working potential is indicative of the $Li_4Ti_5O_{12}$–$Li_7Ti_5O_{12}$ two phase reaction during the lithiation/de-lithiation processes. $Li_4Ti_5O_{12}$ shows a spinel crystal structure with a $Fd\overline{3}m$ space group (Fig. 1) and a lattice parameter $a = 0.836$ nm [3]. All O^{2-} ion occupy the 32e sites, 75% of Li^+ ions occupy the tetrahedral 8a sites, 25% of Li^+ ions and all Ti^{4+} ions with a molar ratio of 1:5 occupy the octahedral 16d sites, while all octahedral 16c sites are empty. Thus, $Li_4Ti_5O_{12}$ can be denoted as $[Li_3]^{8a}[LiTi_5]^{16d}[O_{12}]^{32e}$. The three-dimensional 8a–16c–8a network is identified as the Li^+-ion transport pathways in $Li_4Ti_5O_{12}$. During the lithiation process, the three Li^+ ions at the 8a sites together with three external Li^+ ions from the cathode cooperatively transport to the 16c sites accompanied by the reduction of three Ti^{4+} ions to Ti^{3+} ions, generating $[Li_6]^{16c}[LiTi_5]^{16d}[O_{12}]^{32e}$. This process is reversed during the de-lithiation process. During this electrochemical reaction, the robust three-dimensional framework $[LiTi_5]^{16d}[O_{12}]^{32e}$ is not changed and its unit cell volume change is extremely small (<0.1%) [4]. Such "zero-strain" renders $Li_4Ti_5O_{12}$ with excellent reversibility and cyclic stability.

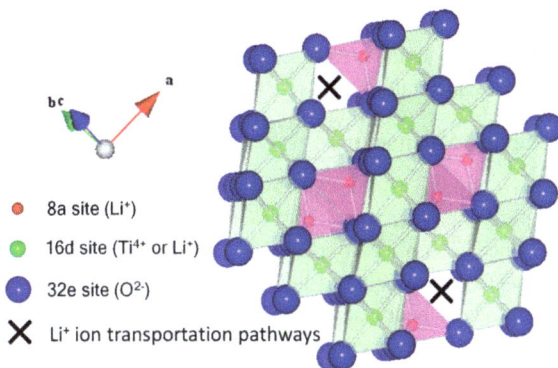

Figure 1. Crystal structure of $Li_4Ti_5O_{12}$. Reproduced from Ref. 5 with permission from the Royal Society of Chemistry.

Despite of the above advantages, $Li_4Ti_5O_{12}$ suffers from three problems. Firstly, it shows poor electronic conductivity and Li^+-ion diffusion coefficient, limiting its rate performance. Secondly, it exhibits a small theoretical capacity of only 175 mAh g^{-1} in 3–1 V, which is significantly smaller than that of graphite. Thus, the LIBs with the $Li_4Ti_5O_{12}$ anodes have low energy density. Finally, severe gassing appears in the $Li_4Ti_5O_{12}$ electrodes, limiting the practical use of $Li_4Ti_5O_{12}$. In this chapter, the improvements of the rate performance, the capacity and the gassing of $Li_4Ti_5O_{12}$ are reviewed. The last part of this chapter provides our proposed future research and development of $Li_4Ti_5O_{12}$.

2. Rate-Performance Improvement

The electronic conductivity and Li^+-ion diffusion coefficient in the $Li_4Ti_5O_{12}$ particles, the electrical conduction between the $Li_4Ti_5O_{12}$ particles, and the $Li_4Ti_5O_{12}$ particle size are the four key factors determining the rate performance of $Li_4Ti_5O_{12}$. Crystal structure modification (including doping), composing and nanosizing are the three main methods to improve the rate performance of $Li_4Ti_5O_{12}$. Crystal structure modification can alter the lattice parameters, thus can alter the Li^+-ion diffusion coefficient. Moreover, this method can introduce some conductive ions with unpaired electrons, thus can improve the electronic conductivity. Composing with a conductive phase can improve the electrical conduction between the $Li_4Ti_5O_{12}$ particles. Nanosizing is a very effective method to improve the rate performance since the transport distance of electrons and Li^+ ions in the $Li_4Ti_5O_{12}$ particles are significantly reduced. Furthermore, the methods combining the above three methods may have better effects on the rate-performance improvement since such combined methods can simultaneously improve the above factors. In addition, a few novel electrode designs can result in good rate performances.

2.1 Crystal Structure Modification

Lin *et al.* studied $Li_{4-2x}Ni_{3x}Ti_{5-x}O_{12}$ ($0 \le x \le 0.15$) anode materials from a solid-state reaction [6]. In the Ni^{2+} doping, the basic spinel crystal structure did not changed, but every two Li^+ ions and every one Ti^{4+} ion were replaced by three Ni^{2+} ions. Since the size of three Ni^{2+} ions (0.67 Å) were almost equal to the total size of two Li^+ ions (0.76 Å) and one Ti^{4+} ion (0.605 Å) in the octahedral sites, the lattice parameters of $Li_{4-2x}Ni_{3x}Ti_{5-x}O_{12}$ were almost equal. Furthermore, there were few non-Li^+-ions occupying the 8a sites after the Ni^{2+} doping. Consequently, the Li^+-ion diffusion coefficient was mostly unchanged. However, the unpaired 3d electrons in the introduced Ni^{2+} ions improved the electronic conductivity by at least one order of magnitude. Due to this improvement, $Li_{3.9}Ni_{0.15}Ti_{4.95}O_{12}$ showed an improved rate performance with a large capacity of 72 mAh g^{-1} at 5 C, which was 118% larger than that of $Li_4Ti_5O_{12}$. Later, Lin *et al.* also investigated Cu^{2+} doped $Li_4Ti_5O_{12}$ [7]. Similar to the Ni^{2+} doping, the Cu^{2+} doping significantly improved the electronic conductivity due to the unpaired 3d electrons in the Cu^{2+} ions. However, the size of the Cu^{2+} ion (0.73 Å) in the octahedral sites was larger than that of the Ni^{2+} ion. Thus, the lattice parameters of $Li_{4-2x}Cu_{3x}Ti_{5-x}O_{12}$ were increased (Fig. 2a), benefiting the Li^+-ion transport. On the other hand, the Cu^{2+} doping rendered a small amount of non-Li^+ ions to occupy the 8a sites, slightly blocking the Li^+-ion transport pathways. In this doping, the former factor carried more weight than the latter one. Therefore, the Li^+-ion diffusion coefficients of $Li_{4-2x}Cu_{3x}Ti_{5-x}O_{12}$ ($x = 0.05$, 0.1, 0.15) were larger than that of $Li_4Ti_5O_{12}$ (Fig. 2b). Due to the improvements of the electronic conductivity and Li^+-ion diffusion coefficient, $Li_{3.8}Cu_{0.3}Ti_{4.9}O_{12}$ exhibited a larger capacity of 78 mAh g^{-1} at 10 C (Fig. 2c). This capacity was ~6 times larger than that of $Li_4Ti_5O_{12}$. Besides the Ni^{2+} and Cu^{2+} doped $Li_4Ti_5O_{12}$ materials, Fe^{2+}, Co^{2+}, Cr^{3+} and Cu^+ doped $Li_4Ti_5O_{12}$ materials were also studied by Lin *et al.* [5,8–10]. They concluded that the dopants with unpaired 3d electrons, large sizes and large octahedral site preference energy are desirable for the rate-performance improvement.

Besides the doping, the non-stoichiometric design can also modify the crystal structure and improve the rate performance. Chen *et al.* calcined the $Li_4Ti_5O_{12}$ precursor from a solvothermal reaction in N_2 to prepare $Li_4Ti_5O_{12-y}$ [11]. Its electron paramagnetic spectra demonstrated the existence of O^{2-} vacancies. Thus, a small amount of Ti^{4+} ions in the original $Li_4Ti_5O_{12}$ were reduced to be Ti^{3+} ions in $Li_4Ti_5O_{12-y}$ to keep the charge balance. The unpaired 3d electrons in the Ti^{3+} ions significantly improved the electronic conductivity of $Li_4Ti_5O_{12-y}$. Furthermore, the existence of O^{2-} vacancies and large Ti^{3+} ions led to the improvement of the Li^+-ion diffusion coefficient. As a result of such non-stoichiometric design, $Li_4Ti_5O_{12-y}$ showed a better rate performance than its stoichiometric part. $Li_4Ti_5O_{12-y}$ delivered a large capacity of 125 mAh g^{-1} at 20 C, whereas that of $Li_4Ti_5O_{12}$ was only 66 mAh g^{-1}.

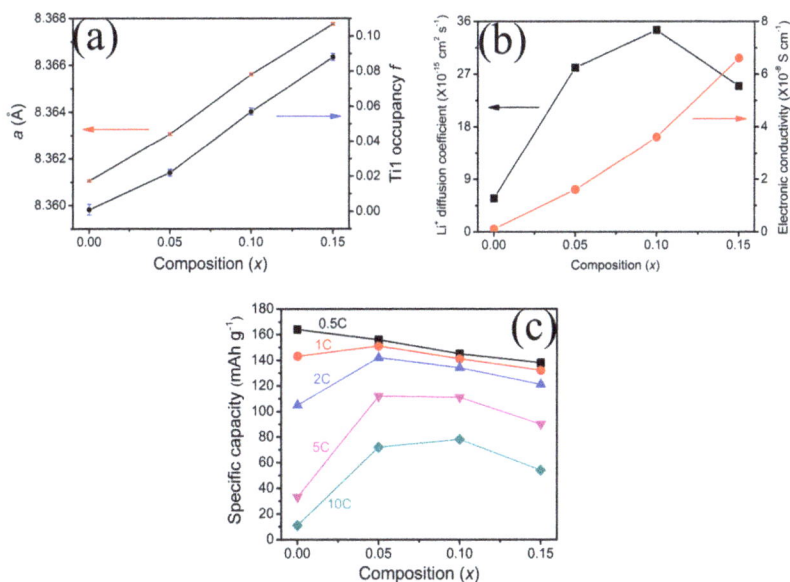

Figure 2. (a) Lattice parameter and occupancy of Ti^{4+} ion in 8a sites, (b) Li^+-ion diffusion coefficient and electronic conductivity, and (c) rate performances of $Li_{4-2x}Cu_{3x}Ti_{5-x}O_{12}$ (x = 0, 0.05, 0.1 and 0.15). Reproduced from Ref. 7 with permission from Elsevier.

2.2 Compositing

Yang *et al.* prepared a graphene-wrapped $Li_4Ti_5O_{12}$ composite. P25 (TiO_2 nanoparticles with a particle size of ~25 nm) and graphene oxide were uniformly mixed by ultrasonication [12]. After drying, the resultant graphene oxide-wrapped P25 nanoparticles were mechanically mixed with Li_2CO_3 at a molar ratio of Li : Ti = 4.1 : 5. Then, the mixture was calcined at 800 °C for 10 h in H_2/Ar to form the graphene-wrapped $Li_4Ti_5O_{12}$ composite. The graphene flakes were well dispersed in the composite, improving the electrical conduction between the $Li_4Ti_5O_{12}$ particles and thus the rate performance. At 10 C, this composite delivered a large capacity of 147 mAh g^{-1}. Krajewski *et al.* synthesized a $Li_4Ti_5O_{12}/Ag$ composite [13]. $Li_4Ti_5O_{12}$ particles from a solid-state reaction were added to an $AgNO_3$ solution. After dried, the mixture was dried at 150 °C for a few hours to form the $Li_4Ti_5O_{12}/Ag$ composite with an Ag-particle size of 2–8 nm. The tiny Ag particles enhanced the electrical conduction between the $Li_4Ti_5O_{12}$ particles. This composite remained ~86% of the capacity when increasing the current rate from 1 C to 10 C.

2.3 Nanosizing

So far, various $Li_4Ti_5O_{12}$ nanomaterials, including zero-dimensional nanoparticles, one-dimensional nanorods/nanotubes/nanowires, two-dimensional nanosheets and three-dimensional porous spheres, have been reported and showed improved rate performances.

Zhang *et al.* applied a glycerol assisted hydrothermal method to fabricate $Li_4Ti_5O_{12}$ nanoparticles [14]. Glycerol controlled the particle size during the hydrothermal process. The glycerol-assisted sample sintered at 500 °C exhibited a small particle size of ~25 nm and a huge specific surface area of 95.93 m^2 g^{-1}, thus delivered excellent electrochemical performances in terms of the rate performance and cyclic stability. At 20 C, it showed a large capacity of 155 mAh g^{-1}. At 5 C, it showed a large capacity retention of 93.3% over 500 cycles. $Li_4Ti_5O_{12}$ nanoparticles can also be synthesized through a supercritical method, as reported by Nugroho *et al.* [15]. Through controlling the feed concentration, reaction time and calcination, the optimized $Li_4Ti_5O_{12}$ displayed good crystallinity and a large specific surface area of 34.2 m^2 g^{-1}. Therefore, it delivered an advanced rate performance with a large capacity of 101 mAh g^{-1} at 10 C. Its electrochemical performances were significantly better than those of the solid-state synthesized $Li_4Ti_5O_{12}$.

Jo *et al.* fabricated $Li_4Ti_5O_{12}$ nanofibers using an electrospinning method with titanium(IV) isopropoxide, lithium acetate dehydrate and poly(vinylpyrrolidone) as the precursors [16]. These nanofibers were highly porous and showed a uniform diameter of <300 nm. Thus, this material exhibited a good rate performance. Its capacities at 0.1 C and 10 C are 160 mAh g^{-1} and 138 mAh g^{-1} (82.3% retained). Zhou *et al.* prepared

$Li_4Ti_5O_{12}$ nanorods through a hydrothermal method using TiO_2-B nanorods as the titanium source and template [17]. These nanorods had a diameter of 100–200 nm and a length of 1–2 μm. At 20 C, this material delivered a capacity of 126 mAh g^{-1} with capacity retention of 80.6% over 1000 cycles.

Li et al. synthesized $Li_4Ti_5O_{12}$ nanosheets by the calcination of $Li_{1.81}H_{0.19}Ti_2O_5 \cdot xH_2O$ nanosheets at 500 °C for 2 h [18]. The nanosheets showed single crystallinity and a thickness range of 10–20 nm. Consequently, this material showed an outstanding rate performance. Its capacity at 10 C was as large as 120 mAh g^{-1} even after 300 cycles. Recently, Sha et al. fabricated $Li_4Ti_5O_{12}$ nanoplates through a hydrothermal method using lithium hydroxide as the lithium source, titanium(IV) butoxide as the titanium source and benzyl alcohol–$NH_3 \cdot H_2O$ as the solvent [19]. After calcined at 500 °C, the nanoplates had good crystallinity. As a result, this material showed excellent electrochemical performances with a reversible capacity of 153 mAh g^{-1} at 10 C even after 1000 cycles.

Lin et al. synthesized monodispersed mesoporous $Li_4Ti_5O_{12}$ submicrospheres through a solvothermal reaction of monodispersed TiO_2 submicrospheres and LiOH in a water–ethanol solvent [20]. After calcined at 600 °C, the optimized $Li_4Ti_5O_{12}$ submicrospheres showed good crystallinity, a sphere size of ~660 nm and a large specific surface area of 15.5 $m^2 g^{-1}$ (Fig. 3a). These submicrospheres were constructed by primary particles with a particle size of 20–100 nm and pores with a pore size of ~4.5 nm. Consequently, this material delivered a stable and large capacity of 109 mAh g^{-1} at 10 C (Fig. 3b). Yu et al. prepared $Li_4Ti_5O_{12}$ hollow submicrospheres through a facile templating method [21]. Firstly, TiO_2 was coated on uniform SiO_2 nanospheres by a sol–gel method. Then, a mixture of $SiO_2@TiO_2$, LiOH and DI water was treated by a hydrothermal reaction. Finally, the product was washed and annealed at 500 °C to form crystallized $Li_4Ti_5O_{12}$ hollow spheres. The uniform hollow spheres had a sphere size of ~600 nm and a shell thickness of ~55 nm. They were composed by $Li_4Ti_5O_{12}$ primary nanoparticles with a particle size of ~9 nm, resulting in a huge specific surface area of 220 $m^2 g^{-1}$. Therefore, this material exhibited an outstanding rate performance. At 1 C, its capacity was 170 mAh g^{-1}. At 20 C, it still remained a large value of 104 mAh g^{-1}. Similarly, Cheng et al. fabricated $Li_4Ti_5O_{12}$ hollow urchin-like submicrospheres through the same templating method but a lower hydrothermal temperature [22]. The microspheres were constructed by $Li_4Ti_5O_{12}$ nanosheets and had a large specific surface area of 140 $m^2 g^{-1}$. As a result, this material showed a large capacity of 120 mAh g^{-1} with small capacity loss of <2% at 10 C.

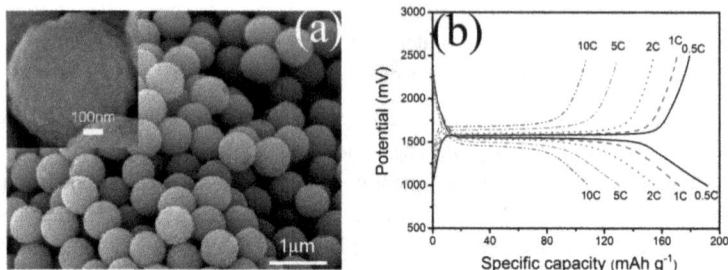

Figure 3. (a) FESEM image and (b) discharge–charge profiles of monodispersed mesoporous $Li_4Ti_5O_{12}$ submicrospheres. Reproduced from Ref. 20 with permission from the Royal Society of Chemistry.

2.4 Combined Methods

Zhang *et al.* prepared graphene oxide coated $Li_4Ti_5O_{12}$ microspheres through a hydrothermal method using LiOH, TiO_2 microspheres and graphene oxide as the precursors [23]. Due to the confined effect of graphene oxide on the surface of TiO_2 microspheres, the size of the resultant $Li_4Ti_5O_{12}$ microspheres was limited to be ~1 µm, These $Li_4Ti_5O_{12}$ microspheres were constructed by $Li_4Ti_5O_{12}$ primary nanoparticles and well-dispersed graphene oxide, which improved the electrical conduction between the $Li_4Ti_5O_{12}$ nanoparticles. Due to the combination of nanosizing and compositing, the 3 wt% graphene oxide coated $Li_4Ti_5O_{12}$ microspheres exhibited an advanced rate performance. At 5 C, this material displayed a large reversible capacity of 132 mAh g^{-1} with 97% capacity retention over 500 cycles. Kim *et al.* fabricated Zr^{4+}-doped $Li_4Ti_5O_{12}$ nanofibers using an electrospinning method [24]. A uniform mixture of zircon(IV) butoxide, titanium(IV) isopropoxide and polyvinylpyrrolidone was dissolved in a solvent and then treated by the electrospinning process. After calcined at 750 °C for 3 h, the resultant nanofibers showed a fiber diameter of ~100 nm. The small diameter led to the small transport distance of electrons and Li$^+$ ions in the nanofibers. In addition, due to the large size of Zr^{4+} ions, the Zr^{4+} doping increased the lattice parameters of $Li_4Ti_5O_{12}$ and thus its Li$^+$-ion diffusion coefficient. Therefore, through combining nanosizing and doping, the Zr^{4+}-doped $Li_4Ti_5O_{12}$ nanofibers revealed an advanced rate performance. The optimized sample delivered a large capacity of 151 mAh g^{-1} at 0.1 C and retained a large value of 121 mAh g^{-1} at 30 C. Lin *et al.* synthesized a $Cu_{0.3}Li_{3.8}Ti_{4.9}O_{12}$/CNTs composite through simply mixing CNTs and $Cu_{0.3}Li_{3.8}Ti_{4.9}O_{12}$ particles from a solid-state reaction [7]. The Cu^{2+} doping improved the electronic conductivity and Li^+-ion diffusion coefficient of $Li_4Ti_5O_{12}$. The mixed CNTs improved the electrical conduction between the $Cu_{0.3}Li_{3.8}Ti_{4.9}O_{12}$ particles. As a result, the $Cu_{0.3}Li_{3.8}Ti_{4.9}O_{12}$/CNTs composite exhibited an outstanding rate performance. Its capacity at 10 C is up to 114 mAh g^{-1}, which is 9.4 times larger than that of $Li_4Ti_5O_{12}$.

Lin *et al.* prepared a nanosized $LiCrTiO_4$/CNTs composite. Li_2CO_3, Cr_2O_3, TiO_2 and CNTs with a predetermined molar ratio were mixed and ball-milled by a high-energy ball-milling machine [8]. The resultant mixture was calcined at 800 °C for 4 h in Ar to form the nanosized $LiCrTiO_4$/CNTs composite (Fig. 4a). The incorporation of CNTs not only hindered the growth of the $LiCrTiO_4$ particles, but also enhanced the electrical conduction between the nanosized $LiCrTiO_4$ particles. The Cr^{3+}-substitution in $Li_4Ti_5O_{12}$ resulted in enhanced electronic conductivity and Li^+-ion diffusion coefficient respectively due to the unpaired 3d electrons in Cr^{3+} ions and the decreased structural disorder. Consequently, the nanosized $LiCrTiO_4$/CNTs composite exhibited an outstanding rate performance with a large capacity of 120 mAh g^{-1} (Fig. 4b), whereas that of $Li_4Ti_5O_{12}$ was only 11 mAh g^{-1} and that of the microsized $LiCrTiO_4$/CNTs composite was 70 mAh g^{-1}. Lin *et al.* also synthesized mesoporous $Li_4Ti_5O_{12-x}$/C submicrospheres [25]. TiO_2 precursor submicrospheres were prepared by a sol–gel method. Then, the as-prepared TiO_2 submicrospheres and LiOH with a stoichiometric ratio were mixed in a sucrose aqueous solution. After drying, the mixture was calcined at 800 °C for 10 h in Ar. The optimized $Li_4Ti_5O_{12-x}$/C submicrospheres showed a graded sphere-size distribution ranging from 100 nm to 600 nm (Fig. 5a), a carbon content of only 1.06 wt%, a large tape density of 1.71 g cm^{-3} and a large specific surface area of 14.6 m^2 g^{-1}. The submicrospheres were constructed by nanosized $Li_4Ti_5O_{12-x}$ primary particles with carbon coatings, carbon nanoparticles and pores with a pore size of ~4.2 nm. The production of Ti^{3+} ions with unpaired 3d electrons and O^{2-} vacancies during the calcination process in Ar increased the electronic conductivity and Li^+-ion diffusion coefficient of $Li_4Ti_5O_{12}$. Through such synergistic method combining nanosizing, compositing and crystal structure modification, this material exhibited an excellent rate performance with a stable capacity of 119 mAh g^{-1} (Fig. 5b), which was 9.8 times larger than that of microsized $Li_4Ti_5O_{12}$ from the traditional solid-state reaction. Ma *et al.* fabricated carbon-encapsulated F^- doped $Li_4Ti_5O_{12}$ and applied it to LIBs [26]. TiO_2 with a ball-in-ball architecture was firstly prepared through a hydrothermal process using TiF_4 as the titanium source. Then, the as-prepared TiO_2, lithium acetate and glucose were mixed in an ethanol solvent. After dried, the mixture were treated by a solid-state reaction process at 800 °C for 6 h in N_2 to form the carbon-encapsulated F^- doped $Li_4Ti_5O_{12}$ with a ball-in-ball architecture. Such unique architecture had a shell thickness of 40–60 nm for the outer and inner spheres and a large specific surface area of 18.2 m^2 g^{-1}. The F$^-$ doping not only improved the integrity of the Li$_4$Ti$_5$O$_{12}$ structure, but also induced the formation of compensating Ti^{3+} ions, improving the electronic conductivity. The nanoscale carbon encapsulation layer of ~3 nm electrically integrated the F$^-$ doped Li$_4$Ti$_5$O$_{12}$ primary particles without obvious capacity penalty. As a result of the doping, compositing and nanosizing, the best-performing composite exhibited an advanced rate performance. At 1 C, it showed a reversible capacity of ~158 mAh g^{-1} with negligible capacity fading over 200 cycles. At 30 C, it still remained a large capacity of 108 mAh g^{-1}.

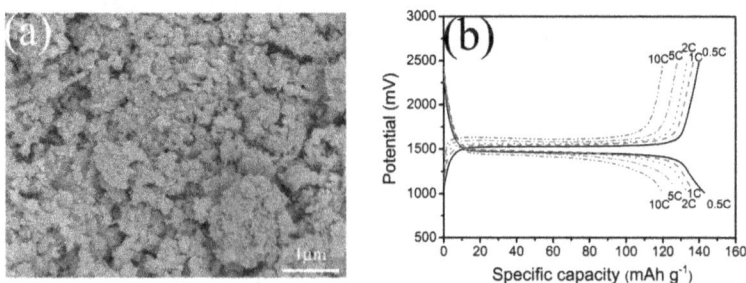

Figure 4. (a) FESEM image and (b) discharge–charge profiles of nanosized LiCrTiO$_4$/CNTs composite. Reproduced from Ref. 8 with permission from the Royal Society of Chemistry.

2.5   Novel Electrode Design

Liu *et al.* prepared self-supported Li$_4$Ti$_5$O$_{12}$/C nanotube arrays on a stainless steel foil and applied this electrode to LIBs [27]. ZnO hexagonal nanorod arrays were firstly grown on a stainless steel foil through a facile and scalable low-temperature solution deposition route. Using the ZnO nanorods as template, TiO$_2$ nanotube arrays were formed on the stainless steel foil through an *in-situ* etching solution route. Then, Li$_4$Ti$_5$O$_{12}$ nanotube arrays were fabricated through a template-based chemical lithiation route using LiOH as the lithium source. Finally, the Li$_4$Ti$_5$O$_{12}$ nanotube arrays were immersed into a glucose aqueous solution and calcined at 550 °C for 5 h in Ar to form self-supported Li$_4$Ti$_5$O$_{12}$/C nanotube arrays on the stainless steel foil. These nanotubes constructed by Li$_4$Ti$_5$O$_{12}$ particles had a diameter of ~250 nm and a wall thickness of ~25 nm. This novel nanotube architecture significantly reduced the transport distance of electrons and Li$^+$ ions in the Li$_4$Ti$_5$O$_{12}$ particles. Moreover, the carbon uniformly coated on the outer and inner surfaces of the $Li_4Ti_5O_{12}$ nanotubes enhanced the electrical conduction between the $Li_4Ti_5O_{12}$ particles. As a result, this flexible and binder-free $Li_4Ti_5O_{12}/C$ nanotube-array electrode showed an excellent rate performance. At 30 C, it delivered a large capacity of 154 mAh g^{-1} with only ~7% capacity loss over 500 cycles.

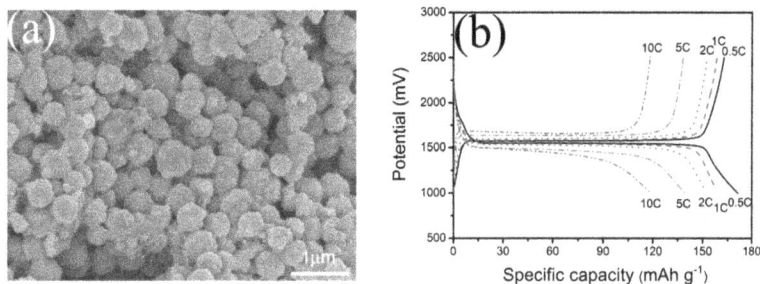

Figure 5. (a) FESEM image and (b) discharge–charge profiles of mesoporous $Li_4Ti_5O_{12-x}/C$ submicrospheres. Reproduced from Ref. 25 with permission from the PCCP Owner Societies.

Shen *et al.* fabricated a three-dimensional $Li_4Ti_5O_{12}/C$ anode [28]. Ultrathin TiO_2 nanosheets were grown on carbon textiles through a solvothermal route. Then, the TiO_2 nanosheet coated carbon textiles were treated by a chemical lithiation process to form the three-dimensional $Li_4Ti_5O_{12}/C$ electrode, in which $Li_4Ti_5O_{12}$ nanoparticles with an average size of ~12 nm were coated on the surface of the carbon textiles. Such novel electrode design exhibited the easy access of Li^+ ions, the decrease of electrons/Li^+-ion transport pathways and the direct electronic pathways to the current collector. As a result, this flexible and binder-free electrode displayed an excellent rate performance with a large capacity of 119 mAh g^{-1} even at 60 C.

Wang *et al.* synthesized a web-like $Li_4Ti_5O_{12}$ anode [29]. A Ti foil was placed in an autoclave filling with a NaOH aqueous solution and treated by a hydrothermal process. Then, the Ti foil covered sample was immersed in an HCl solution to replace Na^+ ions by H^+ ions. $Li_4Ti_5O_{12}$ was formed through a chemical lithiation route using the ion-exchanged sample as the titanium source and LiOH as the lithium source. After calcined at 600 °C for 1.5 h in N_2, crystallized $Li_4Ti_5O_{12}$ on Ti foil was produced, showing a three-dimensional web-like architecture consisting of many nanowires of length ~20 μm. Such novel electrode design led to an outstanding rate performance. At 80 C, this flexible electrode without any carbon or binders delivered a capacity of 103 mAh g^{-1}.

3. Capacity Improvement

$Li_4Ti_5O_{12}$ has a low theoretical capacity of 175 mAh g^{-1} in 3–1 V with three of the five Ti^{4+} ions reduced to Ti^{3+} ions. When decreasing the potential to ~0 V, the other two Ti^{4+} ions are reduced and the additional two external Li^+ ions move to the 8a sites [30], forming $[Li_2]^{8a}[Li_6]^{16c}[LiTi_5]^{16d}[O_{12}]^{32e}$. Thus, $Li_4Ti_5O_{12}$ has an improved capacity of 293 mAh g^{-1} in 3–0 V. However, it is worth noting that the lithiation in 1–0 V is considerably irreversible, leading to a low first-cycle Coulombic efficiency.

For the first time, Lin *et al.* proposed that the capacity of $Li_4Ti_5O_{12}$ could be improved through combining the Ti^{3+}/Ti^{4+} redox couple and other redox couples [31]. They fabricated spinel $Li_5Cr_9Ti_4O_{24}$ as a new anode material (Fig. 6a). $Li_5Cr_9Ti_4O_{24}$ was considered as a modified $Li_4Ti_5O_{12}$ in which some Li^+ and Ti^{4+} ions were replaced by Cr^{3+} ions following $3Cr^{3+} \rightarrow Li^+ + 2Ti^{4+}$. Based on the two redox couples of Ti^{3+}/Ti^{4+} and Cr^{2+}/Cr^{3+}, $Li_5Cr_9Ti_4O_{24}$ exhibited a large theoretical capacity of 323 mAh g^{-1} in 3–0 V, larger than that of $Li_4Ti_5O_{12}$. As a result, $Li_5Cr_9Ti_4O_{24}$ delivered a large first-cycle discharge/charge capacity of 311/220 mAh g^{-1} in 3–0.001 V at 62.5 mA g^{-1} (Fig. 6b). In addition, the Cr^{3+} substitution also improved the rate performance due to the unpaired 3d electrons in the Cr^{3+} ions and the decreased structural disorder. At the huge current density of 4000 mA g^{-1}, its capacity remained a large value of 119 mAh g^{-1}.

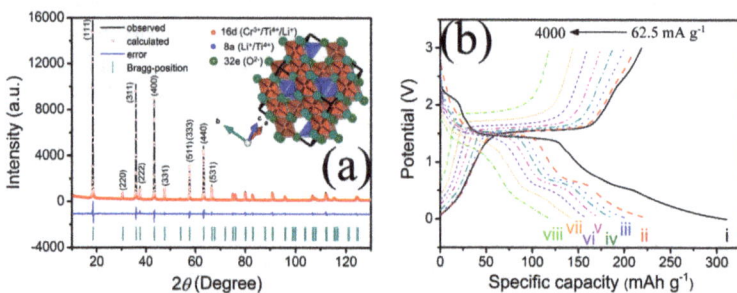

Figure 6. (a) XRD pattern and Rietveld refinement result, and (b) discharge–charge profiles of $Li_5Cr_9Ti_4O_{24}$. Reproduced from Ref. 31 with permission from Elsevier.

4. Gassling Suppression

Gas evolution in LIBs causes irreversible capacity loss and raises serious safety concerns, especially at elevated temperatures. He *et al.* studied the gassing of $Li_4Ti_5O_{12}$ [32]. Using a gas chromatograph, they confirmed that the generated gases mainly contained CO, CO_2

and H_2, which came from the interfacial reactions between the $Li_4Ti_5O_{12}$ particles and the alky carbonate solvents. The reactions occurred at the (111) crystallographic planes of $Li_4Ti_5O_{12}$, leading to the transformation of the (111) planes to (222) planes of $Li_4Ti_5O_{12}$ and the formation of (101) planes of anatase TiO_2. To tackle the gassing problem, carbon was coated on the surface of the $Li_4Ti_5O_{12}$ particles. The carbon coating significantly suppressed the interfacial reactions and thus the gassing. Consequently, carbon-coated $Li_4Ti_5O_{12}$ revealed remarkably improved cyclic stability. This gassing mechanism was confirmed by Guo *et al.* [33]. They proposed that the modification of the crystal facet orientation of $Li_4Ti_5O_{12}$ might be an effective method to suppress the gassing. They fabricated a binder-free $Li_4Ti_5O_{12}$-rutile TiO_2 nanowire array electrode through partially topotactic conversion of rutile TiO_2 nanowire arrays. In this electrode, the outer $Li_4Ti_5O_{12}$ showed a preferred orientation of its (400) planes. Since the exposure of the (111) planes was limited, this hybrid electrode exhibited an ultralong lifetime at 60 °C. After 500 cycles, its capacity at ~2.5 C remained ~83.6% of its initial capacity.

5.   Concluding Remarks and Perspectives

Due to the good safety performance and excellent cyclic stability, $Li_4Ti_5O_{12}$ has drawn much attention in the past 20 years. The poor (electronic and ionic) conductivity was considered as the main disadvantage of $Li_4Ti_5O_{12}$, limiting its rate performance. Thus, intensive studies have been conducted to improve the rate performance. Crystal structure modification can introduce conductive dopants, enlarge the lattice parameters and/or decrease the structural disorder, thus can improve the electronic conductivity and Li^+-ion diffusion coefficient. Compositing can improve the electrical conduction between the $Li_4Ti_5O_{12}$ particles. Nanosizing can decrease the transport distance of electrons/Li^+ ions in the $Li_4Ti_5O_{12}$ particles. These three methods usually significantly improve the rate performance. The combination of these three methods has better effects on the improvements than the single use of either method. Some novel electrode designs can result in good rate performances. The gassing problem of $Li_4Ti_5O_{12}$ can be tackled by surface coating and crystal facet orientation control. While the above two problems can be easily solved, there are no effective methods to significantly increase the capacity up to date. The heavy Cr^{3+} substitution can improve the capacity to a limited extent in 3–0 V. In order to solve the problem of the small capacity, it is highly desirable to develop new $Li_4Ti_5O_{12}$-based materials with the Ti^{3+}/Ti^{4+} redox couple and other redox couples, which may become a hot research topic in near future. If the capacity problem is solved, $Li_4Ti_5O_{12}$ will be a promising and practical anode material for the LIBs for EVs.

References

[1]   M. Armand, J. M. Tarascon: Building better batteries, Nature 451 (2008) 652–657. https://doi.org/10.1038/451652a

[2]   S. S. Zheng: The effect of the charging protocol on the cycle life of a Li-ion batteries, J. Power Sources 161 (2006) 1385–1391. https://doi.org/10.1016/j.jpowsour.2006.06.040

[3]   C. F. Lin, C. Yang, S. W. Lin, J. B. Li: Titanium-containing complex oxides as anode materials for lithium-ion batteries: a review, Mater. Technol. 30 (2015) 192–202. https://doi.org/10.1080/10667857.2015.1107219

[4]   T. Ohzuku, A. Ueda, N. Yamamoto: Zero-strain insertion material of $Li[Li_{1/3}Ti_{5/3}]O_4$ for rechargeable lithium cells, J. Electrochem. Soc. 142 (1995) 1431–1435. https://doi.org/10.1149/1.2048592

[5]   C. F. Lin, X. Y. Fan, Y. L. Xin, F. Q. Cheng, M. O. Lai, H. H. Zhou, L. Lu: $Li_4Ti_5O_{12}$-based anode materials with low working potentials, high rate capabilities and high cyclability for high-power lithium-ion batteries: a synergistic effect of doping, incorporating a conductive phase and reducing particle size, J. Mater. Chem. A 2 (2014) 9982–9993. https://doi.org/10.1039/c4ta01163e

[6]   C. F. Lin, M. O. Lai, L. Lu, H. H. Zhou, Y. L. Xin: Structure and high rate performance of Ni^{2+} doped $Li_4Ti_5O_{12}$ for lithium ion battery, J. Power Sources 244 (2013) 272–279. https://doi.org/10.1016/j.jpowsour.2013.01.056

[7]   C. F. Lin, B. Ding, Y. L. Xin, F. Q. Cheng, M. O. Lai, L. Lu, H. H. Zhou: Advanced electrochemical performance of $Li_4Ti_5O_{12}$-based materials for lithium-ion battery: synergistic effect of doping and compositing, J. Power Sources 248 (2014) 1034–1041. https://doi.org/10.1016/j.jpowsour.2013.09.120

[8]   C. F. Lin, M. O. Lai, L. Lu, H. H. Zhou: Spinel $Li_{4-2x}Co_{3x}Ti_{5-x}O_{12}$ ($0 \leq x \leq 0.5$) for lithium-ion batteries: crystal structures, material properties, and battery performances, J. Phys. Chem. C 118 (2014) 14246–14255. https://doi.org/10.1021/jp504152s

[9]   C. F. Lin, S. F. Song, M. O. Lai, L. Lu: $Li_{3.9}Cu_{0.1}Ti_5O_{12}$/CNTs composite for the anodes of high-power lithium-ion batteries: intrinsic and extrinsic effects, Electrochim. Acta 143 (2014) 29–35. https://doi.org/10.1016/j.electacta.2014.07.122

[10]   C. F. Lin, M. O. Lai, H. H. Zhou, L. Lu: $Li_{3.33}Cu_{1.005}Ti_{4.665}O_{12}$/CuO composite with $P4_332$ space group for Li-ion batteries: synergistic effect of substituting and

compositing, RSC Adv. 4 (2014) 31196–31200.
https://doi.org/10.1039/C4RA03931A

[11]   X. M. Chen, X. F. Guan, L. P. Li, G. S. Li: Defective mesoporous $Li_4Ti_5O_{12-y}$: An advanced anode material with anomalous capacity and cycling stability at a high rate of 20 C, J. Power Sources 210 (2012) 297–302.
https://doi.org/10.1016/j.jpowsour.2012.03.014

[12]   Y. C. Yang, B. H. Qiao, X. M. Yang, L. B. Fang, C. C. Pan, W. X. Song, H. S. Hou, X. B. Li: Lithium titanate tailored by cathodically induced graphene for an ultrafast lithium ion battery, Adv. Funct. Mater. 24 (2014) 4349–4356.
https://doi.org/10.1002/adfm.201304263

[13]   M. Krajewski, M. Michalska, B. Hamankiewicz, D. Ziolkowska, K. P. Korona, J. B. Jasinski, M. Kaminska, L. Lipinska, A. Czerwinski: Li_4Ti_5O12 modified with Ag nanoparticles as an advanced anode material in lithium-ion batteries, J. Power Sources 245 (2014) 764–771. https://doi.org/10.1016/j.jpowsour.2013.07.048

[14]   W. L. Zhang, J. F. Li, Y. B. Guan, Y. Jin, W. T. Zhu, X. Guo, X. P. Qiu: Nano-$Li_4Ti_5O_{12}$ with high rate performance synthesized by a glycerol assisted hydrothermal method, J. Power Sources 243 (2013) 661–667.
https://doi.org/10.1016/j.jpowsour.2013.06.010

[15]   A. Nugroho, S. J. Kim, K. Y. Chung, J. Kim: Synthesis of $Li_4Ti_5O_{12}$ in supercritical water for Li-ion batteries: reaction mechanism and high-rate performance, Electrochim. Acta 78 (2012) 623–632.
https://doi.org/10.1016/j.electacta.2012.06.060

[16]   M. R. Jo, Y. S. Jung, Y. Kang: Tailored $Li_4Ti_5O_{12}$ nanofibers with outstanding kinetics for lithium rechargeable batteries, Nanoscale 4 (2012) 6870–6875.
https://doi.org/10.1039/c2nr31675g

[17]   Q. Zhou, L. Liu, J. L. Tan, Z. C. Yan, Z. F. Huang, X. Y. Wang: Synthesis of lithium titanate nanorods as anode materials for lithium and sodium ion batteries with superior electrochemical performance, J. Power Sources 283 (2015) 243–250.
https://doi.org/10.1016/j.jpowsour.2015.02.061

[18]   N. Li, T. Mei, Y. C. Zhu, L. L. Wang, J. W. Liang, X. Zhang, Y. T. Qian, K. B. Tang: Hydrothermal synthesis of layered $Li_{1.81}H_{0.19}Ti_2O_5 \cdot xH_2O$ nanosheets and their transformation to single-crystalline $Li_4Ti_5O_{12}$ nanosheets as the anode materials for Li-ion batteries, CrystEngComm 14 (2012) 6435–6440.
https://doi.org/10.1039/c2ce25900a

[19]   Y. J. Sha, B. T. Zhao, R. Ran, R. Cai, Z. P. Shao: Synthesis of well-crystallized $Li_4Ti_5O_{12}$ nanoplates for lithium-ion batteries with outstanding rate capability and cycling stability, J. Mater. Chem. A 1 (2013) 13233–13243. https://doi.org/10.1039/c3ta12620j

[20]   C. F. Lin, X. Y. Fan, Y. L. Xin, F. Q. Cheng, M. O. Lai, H. H. Zhou, L. Lu: Monodispersed mesoporous $Li_4Ti_5O_{12}$ submicrospheres as anode materials for lithium-ion batteries: morphology and electrochemical performances, Nanoscale 6 (2014) 6651–6660. https://doi.org/10.1039/c4nr00960f

[21]   L. Yu, H. B. Wu, X. W. Lou: Mesoporous $Li_4Ti_5O_{12}$ hollow spheres with enhanced lithium storage capability, Adv. Mater. 25 (2013) 2296–2300. https://doi.org/10.1002/adma.201204912

[22]   J. Cheng, R. C. Che, C. Y. Liang, J. W. Liu, M. Wang, J. J. Xu: Hierarchical hollow $Li_4Ti_5O_{12}$ urchin-like microspheres with ultra-high specific surface area for high rate lithium ion batteries, Nano Res. 7 (2014) 1043–1053. https://doi.org/10.1007/s12274-014-0467-2

[23]   J. W. Zhang, Y. R. Cai, J. Wu, J. M. Yao: Graphene oxide-confined synthesis of $Li_4Ti_5O_{12}$ microspheres as high-performance anodes for lithium ion batteries, Electrochim. Acta 165 (2015) 422–429. https://doi.org/10.1016/j.electacta.2015.03.016

[24]   J. Kim, M. Park, S. M. Hwang, Y. Heo, T. Liao, Z. Q. Sun, J. H. Park, K. J. Kim, G. Jeong, Y. Kim, J. H. Kim, S. X. Dou: Zr^{4+} Doping in $Li_4Ti_5O_{12}$ anode for lithium-ion batteries open Li+ diffusion paths through structural imperfection, ChemSusChem 7 (2014) 1451–1457. https://doi.org/10.1002/cssc.201301393

[25]   C. F. Lin, M. O. Lai, H. H. Zhou, L. Lu: Mesoporous $Li_4Ti_5O_{12-x}$/C submicrospheres with comprehensively improved electrochemical performances for high-power lithium-ion batteries, Phys. Chem. Chem. Phys. 16 (2014) 24874–24883. https://doi.org/10.1039/C4CP03826F

[26]   Y. Ma, B. Ding, G. Ji, J. Y. Lee: Carbon-encapsulated F⁻ doped $Li_4Ti_5O_{12}$ as a high rate anode material for Li^+ batteries, ACS Nano 7 (2013) 10870–10878. https://doi.org/10.1021/nn404311x

[27]   J. Liu, K. P. Song, P. A. van Aken, J. Maier, Y. Yu: Self-supported $Li_4Ti_5O_{12}$–C nanotube arrays as high-rate and long-life anode materials for flexible Li-ion batteries, Nano Lett. 14 (2014) 2597–2603. https://doi.org/10.1021/nl5004174

[28]   L. F. Shen, B. Ding, P. Nie, G. Z. Cao, X. G. Zhang: Advanced energy-storage architectures composed of spinel lithium metal oxide nanocrystal on carbon textiles, Adv. Energy Mater. 3 (2013) 1484–1489. https://doi.org/10.1002/aenm.201300456

[29]   X. F. Wang, B. Liu, X. J. Hou, Q. F. Wang, W. W. Li, D. Chen, G. Z. Shen: Ultralong-life and high-rate web-like $Li_4Ti_5O_{12}$ anode for high-performance flexible lithium-ion batteries, Nano Res. 7 (2014) 1073–1082. https://doi.org/10.1007/s12274-014-0470-7

[30]   H. Ge, N. Li, D. Y. Li, C. S. Dai, D. L. Wang: Study on the theoretical capacity of spinel lithium titanate induced by low-potential intercalation, J. Phys. Chem. C 113 (2009) 6324–6326. https://doi.org/10.1021/jp9017184

[31]   C. F. Lin, S. J. Deng, H. Shen, G. Z. Wang, Y. F. Li, L. Yu, S. W. Lin, J. B. Li, L. Lu: $Li_5Cr_9Ti_4O_{24}$: a new anode material for lithium-ion batteries, J. Alloys Compd. 650 (2015) 616–621. https://doi.org/10.1016/j.jallcom.2015.08.070

[32]   Y. B. He, B. H. Li, M. Liu, C. Zhang, W. Lv, C. Yang, J. Li, H. D. Du, B. Zhang, Q. H. Yang, J. K. Kim, F. Y. Kang: Gassing in $Li_4Ti_5O_{12}$-based batteries and its remedy, Sci. Rep. 2 (2012) 913. https://doi.org/10.1038/srep00913

[33]   J. L. Guo, W. H. Zuo, Y. J. Cai, S. M. Chen, S. J. Zhang, J. P. Liu: A novel $Li_4Ti_5O_{12}$-based high-performance lithium-ion electrode at elevated temperature, J. Mater. Chem. A 3 (2015) 4938–4944. https://doi.org/10.1039/C4TA05660D

CHAPTER 5

Nanostructured Materials for Lithium Ion Battery Anodes

Yu Chen

College of Physics, Optoelectronics and Energy, Soochow University, P.R. China

chenyu_ny@suda.edu.cn

Abstract

Owing to its high energy densities in terms of both volume and mass, lithium ion battery (LIB) has quickly dominated the energy storage device market since its commercialization in 1991. As the crucial component, anode has a significant effect on the performance of a LIB. New anode materials are urged to meet the requirements of newly emerging and future LIB applications. The extensive study on nanosized materials brings great opportunity for LIB anodes to further improve their performances in terms of capacity, rate capability, and cyclic stability. In this chapter, the effect of nanosized anodes and research progress of nanostructured anodes with different dimensionalities are reviewed.

Keywords

Lithium Ion Battery, Anode, Nanomaterials

Contents

1.   Introduction

As a type of secondary battery, lithium ion battery (LIB) was brought into the market by SONY in 1991. Owing to its higher energy densities compared with traditional battery systems, such as lead acid and Ni-Cd batteries, great attention has been paid on this technology ever since. As shown in Figure 1 [1], the energy density of LIB is at least 2 – 3 times higher than those of conventional rechargeable batteries. Therefore, LIB quickly dominated the market of portable electronics for past two decades. Figure 2 shows the basic structure of a LIB, including two electrodes that can reversibly store lithium and electrolyte that is insulative to electron but permeable to lithium ions. Upon charging, an external electric power source is connected to two electrodes, driving electrons travel from cathode to anode through external circuit. Within the battery, lithium ions travel from cathode to anode through electrolyte to maintain the electrical neutrality. Discharge reverses this process. Electrons travel from anode to cathode through external circuit, supplying electric energy to the connected devices.

Anode is a crucial component of a LIB. Compared with the fast development of new cathode materials, the commercially available anode is still graphite developed by SONY over 20 year ago. However, various newly emerging applications, such as electrical energy storage systems and electrical vehicles (EVs), require future LIB with much higher performances in terms of energy density, power density, cyclic stability, and safety [2-4]. Therefore, graphite anode, with limited capacity and rate capability, is insufficient to meet the requirement for current and future applications. Various new anode materials have been proposed. Based on their reaction mechanism towards lithium, these new materials can be summarized into three major categories, namely intercalaction-based [5], alloying-based [6], and conversion-based [7] anodes.

Currently, the commercial LIBs are based on micrometer-sized electrode materials. Despite of their relative high tap density, the electrochemical performances of these type of LIBs are limited by slow lithium ion diffusion and poor electric conductivity, thus limiting their further applications in advanced applications such as EV. EV is a rapid growing market for LIBs. It is speculated that, for the LIB market, the share of EVs will exceed that of portable electronics in the near future [8]. The three major parameters of a LIB anode, namely capacity, rate capability and cyclic stability, directly relate to cruising distance, acceleration ability, and operational life of EVs. All these parameters are exactly the shortcomings of the EVs currently available in the market. Therefore, the current challenge in developing LIB anode is to find solution to meet these new requirements.

The recent research on nanoscale science brings great opportunities for LIB anodes to further improve their performances [9]. The development of nanostructured anode materials is considered to be promising towards the fast and wide implementation of future electronic devices and equipments. Numerous research groups around the globe are investigating nanostructured LIB anode materials. Industrial effort has also been made by A123 systems and Nexeon on manufacturing LIBs based on nanosized electrodes. However, it is important to clarify the real impact of "nano" on the performance of anode materials. In this review, the advantages and disadvantages of nanosized electrode materials of LIBs will be briefly introduced. Subsequently, the research progress of nanostructured anode materials with different dimensionalities, including 0D, 1D, and 2D, will be reviewed.

2. Advantages and Disadvantages of Nanosized Anode Materials

2.1 Advantages

New lithium storage mechanism

Some materials, such as metal oxides and sulfides, cannot react with lithium reversibly in their bulk forms. However, as their dimensions are reduced to nanosize regime, owing to a much higher surface to volume reaction, these materials become much more active and making the reversible reaction with lithium possible [10].

Enhanced lithium transportation kinetics

The slow lithium diffusion process within anode particle is one of the main obstacles to realize fast charge/discharge. The impact of particle size on the rate capability can be understood by following equation: $t = L^2/D$, where t is the diffusion characteristic time constant, L is the diffusion length, and D is the diffusion constant. By using nanosized anode materials, the lithium transportation path is reduced greatly, thus enhancing the resultant rate capability.

Better structural integrity

In the search of anode materials with high theoretical capacities, one major problem is their large volume variation during the charge/discharge process. In multiple cases, anode particles are particular vulnerable to cracking upon volume expansion. Extensive research have shown that, compared with their bulk counterpart, nanosized anode materials can better accommodate a large volume change during the charge/discharge process, thus enhancing their cyclic stabilities.

2.2 Disadvantages

High surface energy

Due to the large surface area, the surface energy of nanosized anode materials are much higher than their bulk counterparts. Therefore, nanomaterials tend to agglomerate during charge/discharge processes and thus losing their size advantages. Extra consideration must be taken into account to prevent undesired agglomeration.

Low first Coulombic efficiency

The high specific area of nanosized anode materials requires a large amount of lithium to form SEI in the first cycle. Due to the fact that most of the SEI formation is irreversible, the first Coulombic efficiencies for most nanosized anode materials are in the range of 60 - 80% [11]. Such low values hinder the practical application of these nanomaterials.

Low packing density

Owing to the void space existed among particles, nanosized anode materials exhibit poor packing densities, thus restricting the volumetric energy density of the resultant full cell.

3. 0-dimensional (0D) Anode Nanomaterials

3.1 Nanoparticles

As one of the most common form of nanomaterials, nanoparticles have been consider as LIB anode materials in a pioneer work by J-M. Tarascon and co-workers [10]. In their work published in *Nature* in 2000, they discovered that nanoszied transition metal oxide particles were able to reversibly react with lithium, delivering a reversible capacity of about 2 - 3 times of that of graphite. CoO was chosen as an example to study the reaction mechanism in detail. There were two known reaction mechanisms between electrode and lithium: intercalaction and alloying. However, CoO does not have any vacant site in its crystal structure and thus is impossible to be intercalated by lithium. Besides, CoO does not form alloy with lithium either. In their work, with the aid of *in situ* X-ray diffraction (XRD) and *ex situ* selected area electron diffraction (SAED) (shown in Figure 3), the following lithium storage mechanism was confirmed:

$$CoO + 2Li \leftrightarrow Li_2O + Co$$

It was known that the Li_2O was electrochemical inactive and thus its decomposition was highly unfavorable. In this work, the electrode consisted of nanosized particles with a large proportion of atoms lying near or on the surface. The large proportion of active surface atoms of these nanoparticles made the decomposition of Li_2O possible. Furthermore, in this pioneer work, authors already predicted the detrimental effect of nanoparticles agglomeration on their electrochemical performances.

Owing to their high surface energy, anodes consisting of nanoparticles are highly vulnerable to agglomeration during electrochemical cycles. Although there are works on the electrochemical performances of pristine nanoparticles, these works mainly focus on reaction mechanism instead of electrochemical performance. J. Cho and co-workers reported the lithium storage mechanism of MnP nanoparticles for the first time [12]. They have found that LiMnP was firstly formed upon the lithium insertion of MnP, followed by the formation of LiP and LiP_7. On the charge process, LiMnP decomposed and formed LiP and LiP_5 phases. Liang et al. have reported the tough behavior of Ge nanoparticles upon lithiation [13]. Ge forms an alloy with lithium, and is able to deliver a theoretical capacity of 1600 mAh/g with a large volume expansion about 260%. In this work, by using *in situ* TEM, they found that Ge nanoparticles retained their integrity during the charge/discharge process. 0-D nanoparticles with different morphologies have also been widely studied, including nanospindle [14], nanocube [15], nanoflower [16], nanoplate [17], nanobox [18], etc.

Despite of above-mentioned studies, additional decoration is usually applied on 0D nanosized anode materials. Carbon coating is a common method to enhance the overall electrochemical performances. 1) A layer of carbon acts as a buffer layer to constrain the underlying active material, thus accommodating its volume change and preventing cracking. 2) Carbon layer enhances the electronic conductivity of the anode material. 3)

Carbon coating stabilizes solid electrolyte interface (SEI), which is beneficial to the Coulombic efficiency and cyclic stability.

Huang and co-workers applied a layer of carbon coating on the surface of SnO_2 nanoparticles. Thus, the obtained SnO_2@carbon anode reached 852 mAh g^{-1} after 200 cycles at a current density of 100 mA g^{-1}, whereas another anode consisting of pristine SnO_2 nanoparticels experienced fast decay and delivered only 89 mA g^{-1} after 100 cycles at the same testing condition [19]. Wan et al. synthesized Fe_3O_4 nanospindle with a thin layer of carbon on its surface [14]. They reported that the carbon coating was crucial in obtaining stable cyclic behavior.

In addition to the carbonaceous surface protection layer, void space is also introduced into 0-D nanomaterials to decrease the ion diffusion path, enlarge the specific surface area, and buffer the volume expansion. There are two major categories of void-containing 0-D nanomaterials, namely hollow and porous nanomaterials.

3.2   Hollow Nanomaterials

Hollow materials have been widely researched in the field of LIB anode application. In order to fully utilize their large specific surface area, the interior void space of hollow nanomaterials is usually connected with surroundings by holes with different dimensions. Currently, there are four major strategies to fabrication hollow nanomaterials: hard templating, soft templating, sacrificial templating, and template-free methods.

Hard templating

Hard templating is the most widely used method to fabricate hollow materials. The synthetic procedure is quite straightforward. Firstly, hard templates with desired shape and dimensions are synthesized, followed by surface functionalization of hard template. Subsequently, hard templates are coated with desired precursors, which can form a compact shell via various post treatments. Finally, the hard templates are selectively removed by etching, annealing, etc. The obvious advantage of hard templating method is that the shape, dimension, and size distribution of final hollow structure can be easily controlled by those of hard templates. Lou et al. has reported the synthesis of coaxial SnO_2@carbon hollow nanospheres [20]. Silica nanospheres were synthesized as hard templates, which were subsequently coated by SnO_2 and carbon layers. Silica hard templates were removed by NaOH solution to obtain the final products. The secondary electron microscope (SEM) and transmission electron microscope (TEM) images of SnO_2@carbon hollow nanospheres are shown in Figure 4a and 4b, respectively. Figure 4c and 4b demonstrates the electrochemical performances of SnO_2@carbon hollow nanospheres. Specifically, with the aid of hollow structure, SnO_2@carbon delivered stable reversible capacity through 100 cycles at the current density of 0.8 C. Similarly, polystyrene nanspheres were also used as hard templates to fabricate hollow SnO_2 nanostructure [21]. Park et al. applied a layer of carbon coating on surface oxidized Si spheres [22]. By chemically etching the SiO_2 layer, Si nanospheres was trapped in the cavity center of carbon hollow spheres. The void space incorporated in these hollow structure greatly accommodated the large volume changes of active materials during lithium insertion/removal, thus enhancing the cyclic stabilities of these hollow electrode materials.

Soft templating

Despite of its wide implementation, hard templating method has several intrinsic disadvantages, such as the complexity of hard template fabrication and the difficulty in its core removal. In contrast, the soft templating method offers a simpler approach to fabricate hollow structure by using micelles and organic species as templates. In a work reported by Zhang et al., micelles consisting of polyvinyl pyrrolidone (PVP, Mw ~ 58000) were used as soft templates to $ZnMn_2O_4$ hollow spheres [23]. Moreover, as demonstrated by the formation scheme in Figure 5a, with the aid of a thermal driven contraction process, ball-in-ball hollow structure was obtained by a post annealing treatment. Such novel structure is clearly shown in the TEM image in Figure 5b. As shown in Figure 5c, stable reversible capacities above 600 mAh g^{-1} were recorded at a current density of 400 mA g^{-1} for 120 cycles. Xia et al. synthesized hollow MnO/C microsphere anode using a natural biological material of microalgae as the soft template [24]. The abundant carbohydrate contained in the microalgae was used as carbon source that further enhance the resultant electrochemical performance. As a result, a high reversible capacity of 702 mAh g^{-1} was reached at the 50th cycle under a current density of 100 mA g^{-1}.

Sacrificial templating

Sacrificial template acts as a reactant in the formation of hollow structure, which is partially or completely consumed during the shell formation process. Similar to hard and soft templates, sacrificial templates directly control the dimension and cavity size of the hollow structure. Kirkendall effect is often involved in the hollow structure formation via sacrificial templating method. As discovered by E. O. Kirkendall back in 1947, at the interface between two materials, defect formation are prompted by the different diffusion rates of various atoms, eventually leading to the generation of hollow structure. Such effect has been applied repeatedly in fabrication of nanosized hollow structures. Xia et al. obtained hollow Pb@PbS particles owing to the faster diffusion rate of Pb atoms than that of S atoms [25]. Yin et al. reported the synthesis of cobalt sulfide hollow nanocrystals based on a mechanism similar to the Kirkendall effect [26]. Sun et al. observed the formation process of hollow Fe_3O_4 nanoparticles from Fe-Fe_3O_4 core-shell nanopariticels via the Kirkendall effect [27]. Park et al. synthesized hollow SiO_x nanotubes with enhanced lithium storage property via the Kirkendall effect [28]. The morphology evolution was clearly examined by TEM images shown in Figure 6a-d. Moreover, the mechanism of such effect was studied by quantum molecular dynamics calculations (Figure 6e-h). The obtained hollow SiO_x nanotubes demonstrated much better cycling stability compared with solid Si nanowires as shown in Figure 6i. Hollow TiO_2 nanocrystals as LIB anode materials were synthesized by Yu and co-workers [29]. Ge nanotubes were synthesized by Cho and co-workers via the Kirkendall effect. Such hollow structure delivered high capacities (over 1000 mAh g^{-1}) at a high rate of 40000 mA g^{-1} over 400 cycles, demonstrating a superior rate capability [30].

Template-free

Both soft and hard templating methods require additional step of template removal, which complicates the synthesis procedure and may have undesired effect on the hollow structure. In view of these disadvantages, template-free methods to synthesize hollow structure have been developed. Ostwald ripening describes a process that the large particles grow at the expense of smaller ones. Such mechanism refers to a thermodynamically stable status, and was firstly reported by Wilhelm Ostwald in 1896. Many works on the synthesis of hollow materials via a template-free method based on Ostwald ripening process have been reported, where the outer large particles grew on the consumption of interior particles [31-35]. In the field of anode material synthesis, Ostwald ripening is also often applied to obtain hollow active nanostructures.

Chen et al. reported the synthesis of porous hollow Fe_3O_4 beads with an average size of 700 nm based on Ostwald ripening process [36]. As shown in the schematic representation in Figure 7a, smaller oxide particles located at the core were consumed by the outer larger particle during the prolonged hydrothermal process. The hollow nature of such beads was clearly revealed by the SEM images in Figure 7b. A layer of carbon coating was applied on the surface of porous hollow Fe_3O_4 bead to enhance its electrochemical performance. As shown in Figure 7c, by incorporation large amount of void space, carbon coated porous hollow Fe_3O_4 beads demonstrated better electrochemical cycling stability than their solid counterparts. Hollow Co_3O_4 particles by Ostwald ripening were reported with high reversible capacities over 1000 mA h g^{-1} [37].

3.3 Porous Nanoparticles

A porous structure has been widely applied in the fabrication of anode materials regardless of their dimensions. The reason for introducing porosity is similar to that of the hollow structure. By incorporating appropriate amount of porosity, the solid state diffusion paths of ions/electrons are reduced and the volume expansion of active materials can be better buffered. Various methods can be used to generate porosity within 0D nanoparticles. In this section, three commonly used techniques will be discussed, including Ostwald ripening, soft-templating, and metal-organic framework (MOF).

Ostwald ripening

Ostwald ripening has been introduced in the previous section and is known to be very useful to generate hollow structures. As reported by Chen et al., by further elongate the hydrothermal reaction duration, the smaller nanoparticles in the shell can also be consumed by surrounding larger particles, thus resulting a porous shell structure. Compared with their hollow counterpart, porous hollow Fe_3O_4 nanobeads can better buffer the volume changes and further enhance the electrolyte penetration [38].

Soft-templating

Soft-templating is also often applied in the synthesis of mesoporous carbon-based anodes. Zhao et al. have reported the synthesis of mesoporous carbon spheres with ordered mesopores with diameters of ~ 3 nm, using Pluronic F127 as soft template and phenolic resol as carbon source, respectively [39]. On the basis of such structure, Chen et al. incorporated electrochemically active Fe_3O_4 nanoparticle within the mesoporous carbon nanospheres [40]. The schematic illustration and TEM images are shown in Figure 8a and 8b, respectively. Owing to the fast ion and electron transfer channels (carbon framework and interconnected mesopores, respectively) being directly in contact with lithium storage sites, namely Fe_3O_4 nanoparticles, excellent rate capability was achieved by such composite structure of Fe_3O_4 nanoparticles embedded in mesoporous carbon spheres. Specifically, as shown in Figure 8c, such composite anode delivered over 270 mAh g^{-1} under a high current density of 10000 mA g^{-1}, demonstrating superior high rate performances.

Metal-organic framework assisted synthesis

Metal-organic framework (MOF) is a category of porous crystalline material with a large surface area and high pore volume. Traditionally, owing to their small pore sizes (mostly below 1 nm), MOFs have been widely investigated in the field of gas storage and separation [41]. To explore new application including energy storage, MOFs with larger pore size (mesopore) were developed by new synthetic techniques [42, 43]. Wu et al. synthesized porous CuO hollow octahedral by applying a post thermal treatment of Cu-based MOF as precursor [44]. The porous nature of CuO octahedral can be clearly identified by the SEM and TEM images shown in Figure 9a and 9b, respectively. As shown in Figure 9C, the resultant porous anode materials delivered stable capacity around 500 mAh g^{-1} at a current density of 100 mA g^{-1} over 100 cycles.

4.   1-dimensional(1D) Anode Nanomaterials

One-dimensional (1D) nanomaterials have been widely applied in the field of energy storage applications [45]. As active anode materials, 1D nanomaterials possess several advantages: 1) fast electron conduction along the 1D structure; 2) short ion diffusion path along radial direction; 3) high interfacial area with electrolyte; 4) large void space between individual 1D structures. Therefore, various 1D nanostructures, such as nanorods, nanowires, nanofibers, and nanotube, have been fabricated via both physical and chemical methods. In this section, 1D anode materials will be discussed in terms of their morphologies, including 1D nanostructures, 1D nanostructure with branches, and 1D nanostructure arrays.

4.1   1D Nanostructures

Yu et al. fabricated Ge/C nanowires via calcination of inorganic-organic hybrid precursors [46]. The high theoretical capacity of Ge was stabilized by the carbon content. Therefore, the Ge/C composite nanowires delivered a high initial capacity of 1428 mAh g^{-1} at 0.2C. With the aid of the buffering carbon content, a high reversible capacity of 1200 mAh g^{-1} was remained at 50 cycles, indicating a good cyclic stability of Ge/C composite nanowires. Chen et al. synthesized porous Si nanowires by metal assisted chemical etching of cheap metallurgical silicon [47]. The synthetic scheme was shown in Figure 10a. As revealed by the TEM images in Figure 10b and 10c, a layer of oxide was applied to the porous nanowires simply by a post annealing process to stabilize the cycling behavior during the charge/discharge process. As a result, as shown in Figure 10d, the porous Si/SiO$_x$ nanowires showed an average capacity decay of only 0.04% per cycle during a cycling test of 560 cycles at a current density of 600 mA g^{-1}. Lu et al. fabricated Ni$_2$P nanotube based on Kirkendall effect [48]. Ni nanowires were firstly synthesized as templates. Subsequently, trioctylphosphine (TOP) was introduced as P source and form a Ni-TOP complex. Finally, Ni$_2$P nanotubes were obtained with the assistance of the Kirkendall effect. Compared with Ni$_2$P nanoparticles, the as-synthesized Ni$_2$P nanotubes delivered a much higher initial capacity and an improved cyclic stability. Specifically, Ni$_2$P nanotubes delivered 533 mAh g^{-1} at 0.1C after 50 cycles, whereas Ni$_2$P nanoparticles delivered 296 mAh g^{-1} at the same testing condition. Carbon-doped Co$_3$O$_4$ nanofibers were synthesized by using carbon fiber as both hard template and carbon source [49]. Such carbon doped Co$_3$O$_4$ nanofibers delivered a reversible capacity of 1121 mAh g^{-1} at the 100th cycle under a current density of 200 mA g^{-1}.

4.2   1D Nanostructure with Branches

Branched structures consisting of two different kinds of nanorods, where one acts as backbone and the other acts as branches, have recently been studied. In the field of LIB anode applications, the hierarchical structure and synergistic effect between two different electrochemical active materials of such branched structure are believed to be beneficial to its lithium storage properties. Fan et al. reported a branched hybrid structure where SnO_2 nanowire acted as backbone and α-Fe_2O_3 nanorods acted as branches [50]. As shown in Figure 11a, a six-fold symmetry was obtained owing to the different preferential growth directions induced by the distinct crystallographic planes of SnO_2 backbone. Compared with SnO_2 and α-Fe_2O_3 anodes (Figure 11b), the composite structure demonstrated the lowest initial irreversible capacity and highest initial reversible capacity. As discussed in this work, the Fe nanoparticles generated in the charge/discharge process of α-Fe_2O_3 may improve the reversibility of the conversion step of SnO_2, thus improving the reversible capacity of SnO_2/α-Fe_2O_3. Qian et al. reported the synthesis of hybrid nanostructure consisting of β-MnO_2 stem and α-Fe_2O_3 branches [51]. FeOOH nanrods were firstly grown epitaxially on β-MnO_2 nanorods. Subsequently, α-Fe_2O_3 branches were obtained through a high temperature annealing process. Owing to the synergistic effect, β-MnO_2/α-Fe_2O_3 hybrid structure demonstrated much better cyclic stability compared with either of its component, delivering a reversible capacity of 1028 mAh g^{-1} after 200 cycles at a current density of 1000 mA g^{-1}. Lou et al. also reported a heterostructure where both the inner and outer surface of TiO_2 nanotube were grafted with FeOOH spindles, delivering a high area capacity of over 600 mA h cm^{-2} at a current density of 100 mA cm^{-2} for 50 cycles [52].

4.3   1D Nanostructure Arrays

Additives, including polymeric binder and conductive carbon, are often employed in the electrode fabrication for most of nanostructured anode materials, creating extra interfaces and lowering the real capacity of anode. By directly growing 1D nanostructure on conducting substrate, the need of additives can be eliminated. Besides, the array structure offers excellent conductive channels for every individual 1D nanostructure and improves the structural integrity of anode significantly. However, the major drawback for 1D nanostructure array is the low loading density owing to the nano to micro-scale thickness of active electrode materials. Liu et al. demonstrated the synthesis of SnO_2 nanorods directly grown on flexible alloy substrate [53]. The top- and side-view SEM images of SnO_2 nanorod array are shown in Figure 12a and 12b. As shown in Figure 12c, compared with disorder SnO_2 nanorods and SnO_2 nanoparticles, SnO_2 arrays showed much better cyclic stability owing to their ordered array structure. Manthiram et al. reported electrochemical performances of mesoporous TiO_2-Sn/C core-shell nanowire arrays on Ti foil [54]. A high initial capacity of 769 mAh g^{-1} was achieved by TiO_2-Sn/C nanowire array. Besides, stable cyclic performance was demonstrated by TiO_2-Sn/C nanorod array at various current densities. Guo et al. reported Cu-Si nanoarray directly anchored on current collector [55]. Furthermore, a layer of Al_2O_3 was deposited on the surface of the nanoarray to promote the formation of a stable SEI layer. As a result, the Cu-Si-Al_2O_3 nanoarry delivered a high initial reversible capacity of 1820 mAh g^{-1} at a current density of 300 mA g^{-1}. Under a high current density of 1400 mA g^{-1}, Cu-Si-Al_2O_3 nanoarray still delivered a capacity of 1560 mAh g^{-1} after 100 cycles, demonstrating an excellent cyclic stability.

5. 2-dimensional (2D) Anode Nanomaterials

2D nanomaterial is a category of material that only one dimension is restricted to the size of a few nanometers. 2D nanomaterials possess a large surface area and small size in one dimension, thus enlarging the ion flux across the electrode/electrolyte interface and shortening the ion diffusion path across the restricted dimension, respectively. Besides, the volume change can also be buffered during charge/discharge for 2D nanomaterials. Compared with 0D and 1D nanomaterials, 2D nanomaterials usually demonstrate better electron transportation within the 2D structure, where nanomaterials with lower dimensionalities possess more interfaces.

5.1 Pristine 2D Nanomaterials

Graphene and its derivatives

Graphene is a carbon-based 2D nanostructure with carbon atoms arranging in honeycomb lattice. Owing to its peculiar electronic property, excellent mechanical property, and good chemical stability, graphene has drawn great attention in various applications. In the field of energy storage, graphene possesses a theoretical capacity of 744 mAh g^{-1}, which is two times of that of a commercial available anode, graphite. Besides, the excellent electronic and mechanical properties of graphene are also beneficial to the resultant electrochemical performances.

Several methods have been reported to synthesis defect-free graphene flakes. In the Noble-winning paper by A. K. Geim and K. S. Novoselov, single layer graphene with high quality (Figure 13) was obtained by repeatedly peeling a piece of graphite using a adhesive tape until the thinnest flakes were found [56]. Single layer graphene was also obtained by epitaxial growth on metal surface using the chemical vapor deposition (CVD) method [57, 58]. The major drawback for these two techniques is the very low production yield, thus being unsuitable for battery application. Loh et al. reported a facile method to achieve relative large production yield [59]. By electrochemical charging of graphite electrode in a Li^+/propylene carbonate electrolyte followed by sonication, few-layer graphene flakes were easily obtained.

In the application of batteries, derivatives of graphene, namely graphene oxide (GO) and reduced graphene oxide (RGO), are more often used. GO was firstly reported back in 1859 by B. C. Brodie, where he treated graphite with potassium chlorate and fuming nitric acid [60]. In 1958, W. S. Hummers and R. E. Offeman obtained GO using sodium nitrate, sulfuric acid, and potassium permanganate [61]. Such method is still widely adopted by current researchers. The formation mechanism of GO was studied by A. M. Dimiev and J. M. Tour [62]. In their paper, three distinct steps, including conversion of graphite into graphite intercalation compound (GIC), followed by oxidation of GIC into pristine graphite oxide (PGO), and finally exfoliation of PGO into GO, were identified. GO is suitable for large scale fabrication. Besides, abundant oxygen-containing functional groups exist on GO surface, making it hydrophilic and facilitating surface functionalization. By treating GO with an additional reducing process, RGO can be obtained. In such a process (mainly chemical or thermal reductions), the properties and structure of graphene are partially restored [63].

The electrochemical properties of monolayer and bilayer of graphene was examined and reported by A. W. Dryfe and co-workers [64]. They found that the defects in monolayer has little effect on the voltammetric response of samples. Pan et al. studied the lithium storage properties of various GO [65]. They found that the initial reversible capacity of GO was directly related to the degree of defects within the structure. They reported that edge and internal defects can reversibly store lithium and played predominate roles in lithium storage. The first reversible capacities of electron-beam-reduced GO (with SEM images shown in Figure 14a and 14b) reached 1054 mAh g^{-1}. Such value was much higher than those of theoretical capacities of graphite and graphene. The electrochemical cycling behaviors of graphite, pristine GO, and GO reduced by various methods are shown in Figure 14c.

To enhance the electrochemical performances of graphene-based anodes, heteroatoms were purposely introduced. Cheng et al. showed that nitrogen- and boron-doped graphene delivered high initial capacities of 1043 and 1549 mAh g^{-1}, respectively [66]. The XPS spectra of N-doped and B-doped graphene were shown in Figure 15a and 15b, with corresponding electrochemical cycling performances shown in Figure 15c and 15d, respectively. The heteroatomic defects were responsible for the increased intersheet distance and improved conductivities, thus being crucial to the enhanced electrochemical performances. Phosphorus-doped graphene with improved electrochemical performances compared with those of graphene has also been reported by Hou et al. [67]. Specifically, phosphorus-doped graphene delivered a reversible capacity of 460 mAh g^{-1} with no loss over 80 cycles, whereas the reversible capacity of graphene was reported to be around 280 mAh g^{-1}.

Holes in the graphene basal plane were also purposely introduced to enhance its lithium storage property. Holey graphene can easily remain exfoliated and be penetrated by electrolyte. Furthermore, the additional graphene edge introduced in holey graphene was demonstrated with higher electrochemical activity than those in basal plane. Upon nitrogen doping, the N-doped holey graphene delivered a reversible capacity of 605 mAh g^{-1} at 4000th cycle, demonstrating a superior electrochemical performance [68].

Other 2D nanomaterials

Other electrochemical active materials were also synthesized in 2D morphologies as LIB anodes. Chen et al. reported the synthesize of MoS_2 nanosheets with lateral dimension of several hundred nanometers and thickness around 3 nm via a hydrothermal process [69]. The obtained MoS_2 nanosheet demonstrated a high initial capacity of 1122 mAh g^{-1} at a current density of 100 mA g^{-1}. Li et al demonstrated the synthesis of $Li_4Ti_5O_{12}$ nanosheets with N-doped carbon coating [70]. Such 2D nanomaterials showed excellent rate capability with stable capacity around 150 mAh g^{-1} for 100 cycles under a high rate of 10C.

5.2 Decorated 2D Nanomaterials

Owing to their unique 2D structure and excellent electrical conductivities, graphene and its derivatives have been widely utilized as substrates for electrochemical active nanomaterials with lower dimensionalities (0D and 1D). Such composite structure has multiple advantages, including effective preventions of the agglomeration of nanoparticles and the restacking of 2D substrates, accommodation of large volume change, facilitation of the electrolyte penetration, and enhancement of overall electron conductivity. Owing to the synergistic effect, the composites between graphene and 0D/1D nanomaterials usually demonstrate enhancements in terms of capacity, rate capability, and cyclic stability. Various electrochemically active materials have been decorated on the surface of GO via the oxygen-containing functional groups. A post annealing process is often required to partially restore the properties of graphene by reducing GO to RGO.

Owing to their high theoretical capacity and relative ease in synthesis, metal oxide nanomaterials have been extensively combined with graphene and its derivatives for LIB applications. Cheng et al. reported a well-organized flexible interleaved composite between graphene nanosheet (GNS) and Fe_3O_4 particles with morphology shown in Figure 16a [71]. An initial charge capacity of 900 mAh g^{-1} was recorded at a current density of 35 mA g^{-1}. Interestingly, as shown in the cycling test in Figure 16b, a continuous increment in reversible capacity was observed for GNS/Fe_3O_4 composite over 30 cycles. Such phenomenon was also reported by Chen et al. in their study of the anode application of hollow porous Fe_3O_4 nanobeads/RGO composite [38]. Such phenomenon can be ascribed to reversible growth of a polymeric gel-like film from electrolyte degradation. To further enhance the electrochemical performance of Fe_3O_4/G composite, ultrasmall Fe_3O_4 (USIO) nanoparticles (less than 5 nm) were uniformly decorated on the surface of graphene [72]. USIO/G demonstrated a high initial reversible capacity of 1177 mAh g^{-1} and superior stable cycling ability over 1200 cycles at a current density of 1800 mA g^{-1}. The ultrasmall size of active metal oxide nanoparticles was crucial to the satisfied electrochemical performance by significantly shortening the ion diffusion path length and effectively accommodation of volume change of electrochemical active materials. The effect of particle size was confirmed by the lithium storage property of ultrasmall SnO_2 and graphene composite [73]. Guo et al. reported the synthesis of SnO_2 nanocrystal decorated on N-doped graphene with initial reversible capacity of 1021 mAh g^{-1} and increasing capacity over 500 cycles at a current density of 500 mA g^{-1} [74]. The corresponding morphological and electrochemical characterizations were shown in Figure 17. It was claimed that the Sn-N bonding between SnO_2 nanocrystal and graphene nanosheet effectively pinned the Sn nanoparticles formed in cycling process, thus preventing them from agglomeration. Other electrochemically active nanomaterials, including Co_3O_4 [75], CoO [76], MnO_2 [77], Mn_3O_4 [78], TiO_2 [79], MoS_2 [80], SnS_2 [81], and Sn [82], have also been decorated on graphene and its derivatives to form composite with improved electrochemical performances.

Among all potential anode materials, Si attracts most attention with an outstanding theoretical capacity of 4200 mAh g^{-1} (forming $Li_{4.4}Si$ at full lithiation state). However, large volume expansion (over 300%) hinders it from real application. Therefore, graphene, with unique 2D structure and good mechanical property, is a suitable candidate to accommodate the volume change of Si. Kung et al. reported the electrochemical performance of nanosilicon/RGO composite with relative stable capacities over 150 cycles under various current densities [83]. Chou et al. also demonstrated that with the incorporation of graphene, Si nanoparticles showed improvement in cyclic stability [84]. Zhang et al. demonstrated the synthesis of a polycrystalline graphene hollow sphere isolated Si nanoparticle (Si@void@graphene) composite via an in situ pyrolysis and metal catalyzed graphitization reaction [85]. Si@void@graphene delivered a high initial capacity of 2256 mAh g^{-1} and remained 62.5% of this value after 600 cycles, demonstrating a good cyclic stability. Chen and co-workers fabricated multilayered Si/RGO nanostructure as LIB anode [86]. The fabrication scheme was shown in Figure 18a. The layer-by-layer structure shown in Figure 18b was beneficial to the resultant electrochemical performance by effectively preventing the agglomeration of Si nanoparticles, accommodating the volume change of Si, and creating 3D transportation networks for both electron and lithium. Therefore, as shown in Figure 18c, the Si/RGO anode still delivered a reversible capacity of 765 mAh g^{-1} after 300 charge/discharge cycles at a high rate of 3C without significant fading.

6. Conclusions

Nanomaterials have been proven to be promising candidates for future LIB anode application. Furthermore, the dimensionality of electrochemical active nanomaterials plays an crucial role in the resultant electrochemical performances of anode materials, including capacity, rate capability, and cyclic stability. However, it must be emphasized that nanosized anode materials also have several disadvantages, including high surface energy, low first Coulombic efficiency, and low tap density. Besides, the theory and mechanism underlying nanosized anode materials need to be further investigated by researchers from various disciplines. Upon solving these challenges, the development of next generation energy storage devices can be greatly boosted by nanosized LIB anode materials with different dimensionalities.

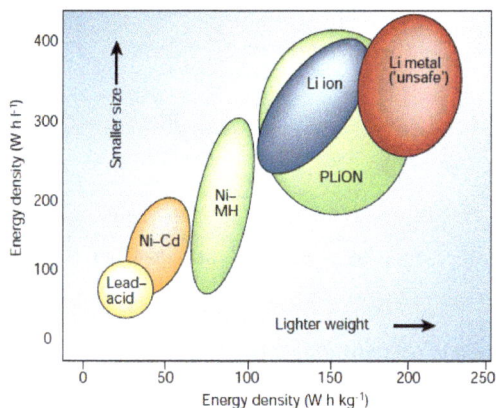

Figure 1. Comparison of the volumetric and gravimetric energy density of different battery technology. Reprinted by permission from Macmillan Publishers Ltd: NATURE ref. [1], copyright 2001.

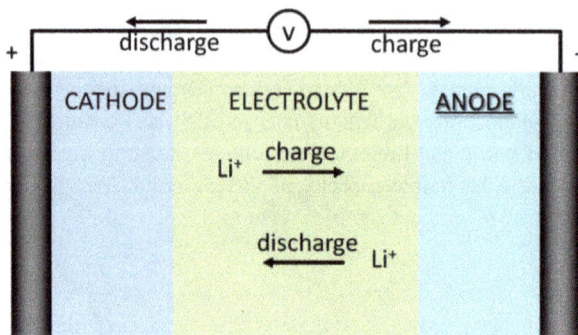

Figure 2. Schematic representation of the structure of a LIB.

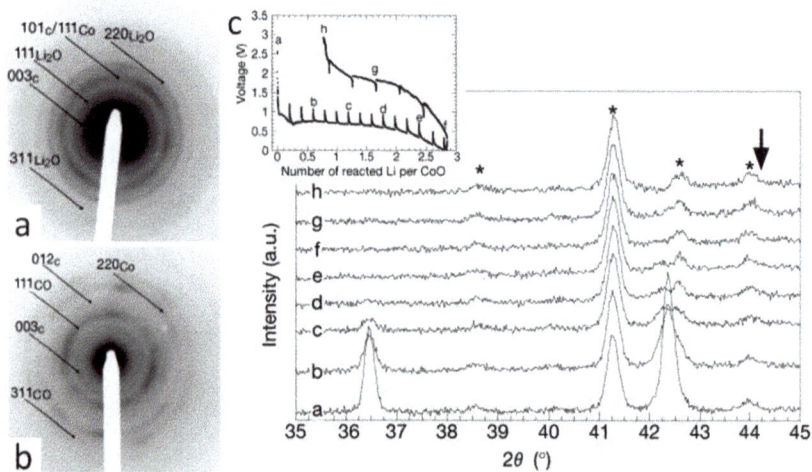

Figure 3. SAED pattern of (a) fully lithiated and (b) delithiated CoO electrode. (c) In situ XRD patterns at different voltages during a charge/discharge process of CoO electrode. Reprinted by permission from Macmillan Publishers Ltd: NATURE ref. [10], copyright 2000.

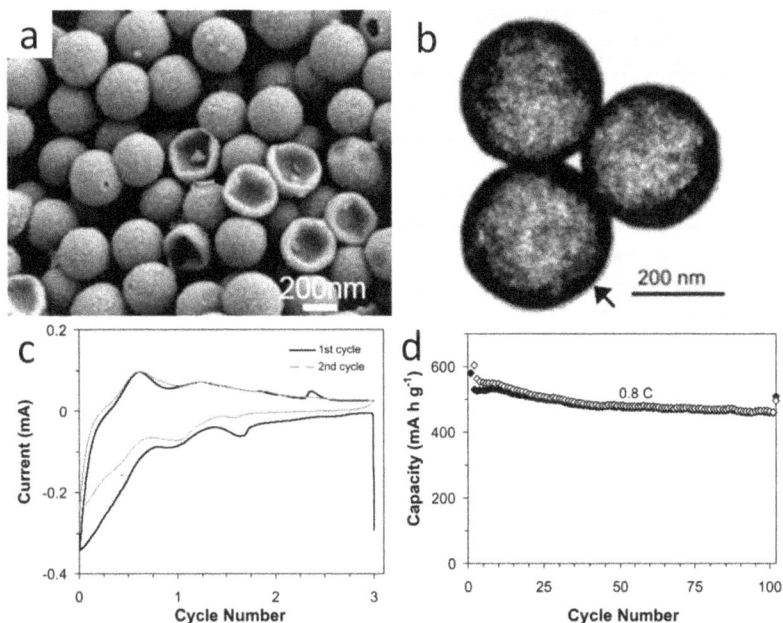

Figure 4.(a) SEM and (b) TEM images of SnO$_2$@carbon hollow spheres.(c) cyclic voltammograms of SnO$_2$@carbon for the first two cycles at a scan rate of 0.05 mV s^{-1} between 3 and 0.005 V. (d)Cycling performance of SnO$_2$@carbon at a scan rate of 0.8 C(C=625 mA g^{-1}). Reprinted with permission from ref. [20]. Copyright 2009 Wiley-VCH Verlag GmbH & Co. KGaA, Weinheim.

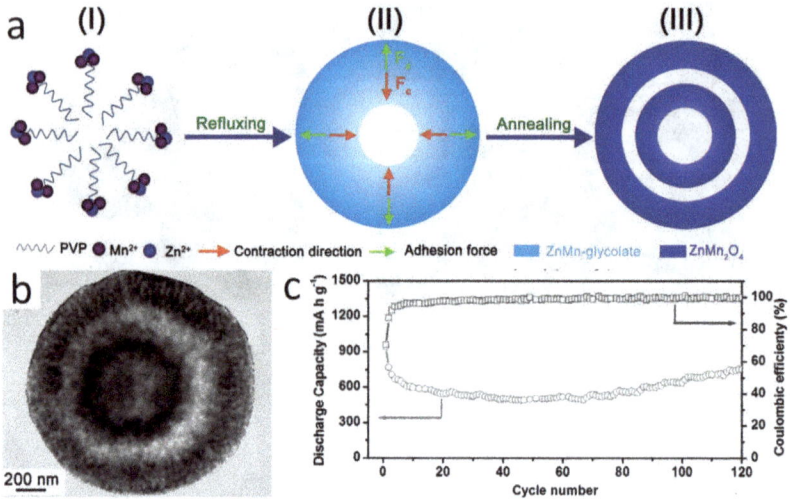

Figure 5. (a) Schematic representation of the formation process of ball-in-ball ZnMn$_2$O$_4$ hollow structure. (b) TEM image and (c) cyclic performance of ZnMn$_2$O$_4$ hollow microspheres. Reprinted with permission from ref. [23]. Copyright 2012 Wiley-VCH Verlag GmbH & Co. KGaA, Weinheim.

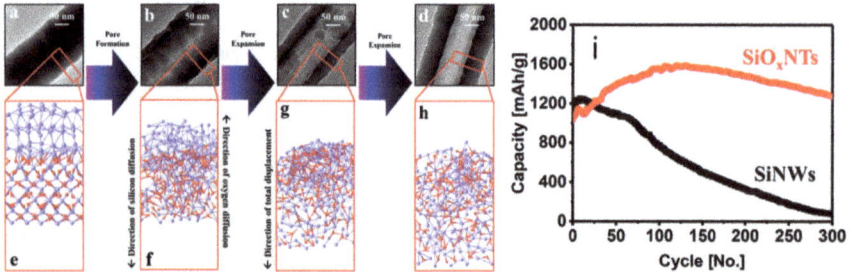

Figure 6. (a-d) TEM images of the morphology evolution from Si nanowire to SiO$_x$ nanotube. (e-h) Simulated representation of the Kirkendall effect of interior void formation of SiO$_x$ nanotube. (i) Electrochemical cycling performance of SiO$_x$ nanotubes and Si nanowires at a current density of 1C. Reprinted with permission from ref. [28]. Copyright 2015 American Chemical Society.

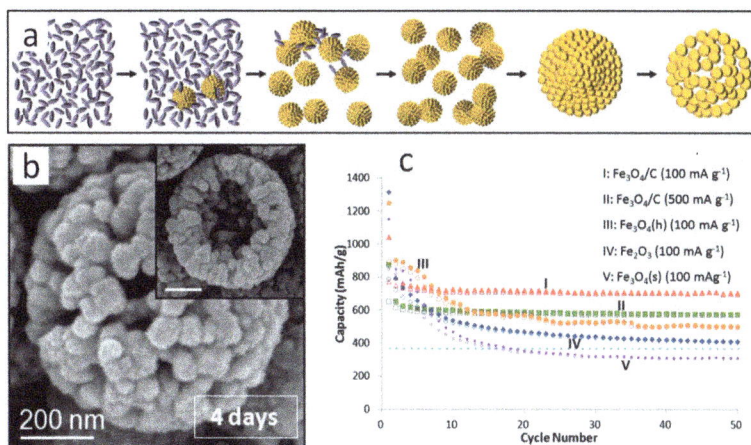

Figure 7. (a) Schematic representation of the Ostwald ripening process of hollow porous Fe_3O_4 beads. (b) SEM images of hollow porous Fe_3O_4 beads. Scale bar in inset: 200 nm. (c) Cycling performances of Fe_3O_4/C beads, hollow Fe_3O_4 beads, solid $Fe3O_4$ beads and hollow Fe_2O_3 beads. Reproduced from Ref. [36] with permission from The Royal Society of Chemistry.

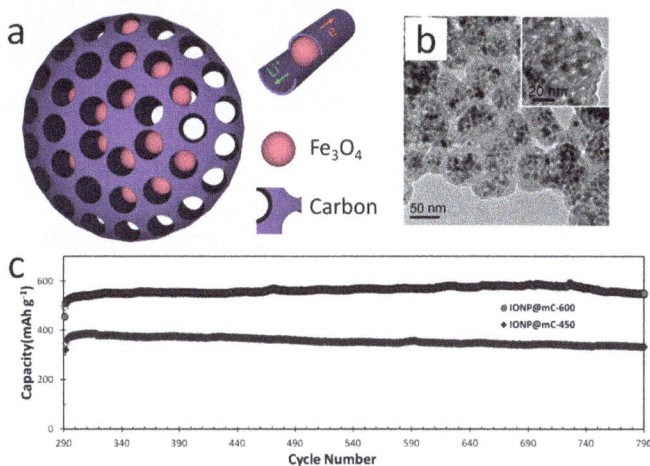

Figure 8. (a) Schematic representation, (b) TEM images, and (c) cycling performance of Fe_3O_4 nanoparticles embedded in mesoporous carbon spheres. Reprinted with permission from ref. [40]. Copyright 2012 Wiley-VCH Verlag GmbH & Co. KGaA, Weinheim.

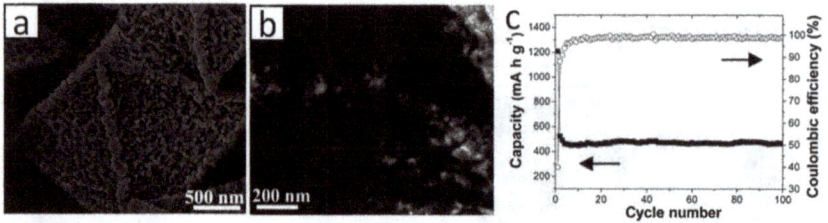

Figure 9. (a)SEM, (b) TEM images, and (c) cyclic behavior of porous CuO hollow octahedral. Reproduced from Ref. [44] with permission from The Royal Society of Chemistry.

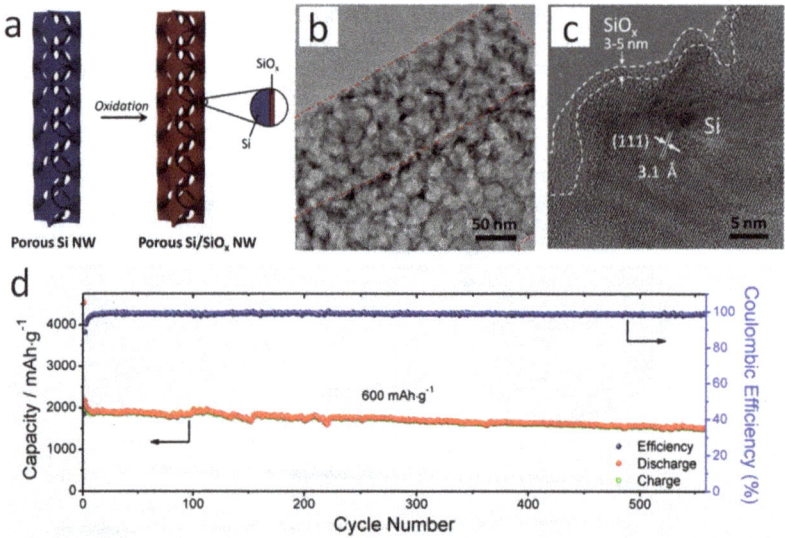

Figure 10. (a)Schematic representation of the formation of porous Si/SiOₓ nanowires. (b-c) TEM images and (d) cycling behavior of porous Si/SiOₓ nanowires. Reprinted with permission from ref. [47]. Copyright 2012 Wiley-VCH Verlag GmbH & Co. KGaA, Weinheim.

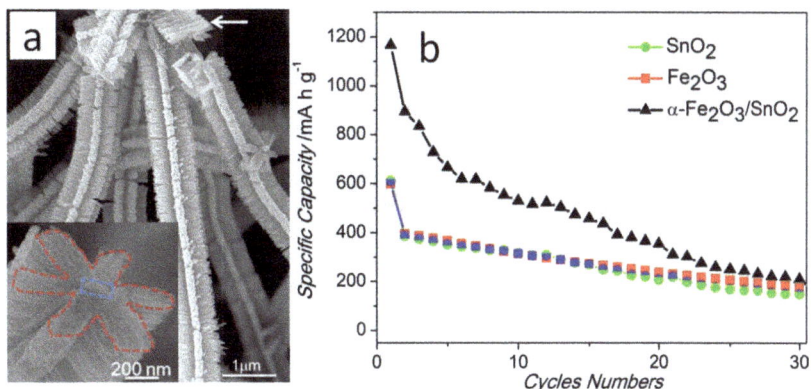

Figure 11. (a) SEM image of α-Fe₂O₃/SnO₂ nanoheterostructure. (b) Electrochemical cycling performance comparison of SnO₂ nanowires, Fe₂O₃ nanorods, and α-Fe₂O₃/SnO₂ nanoheterostructure. Reprinted with permission from ref. [50]. Copyright 2012 Wiley-VCH Verlag GmbH & Co. KGaA, Weinheim.

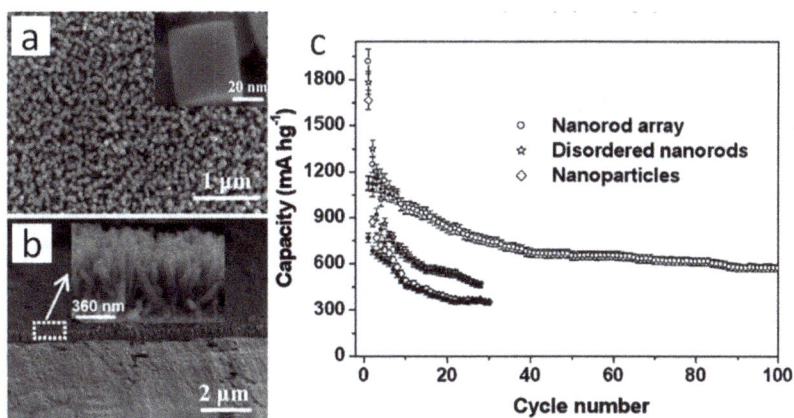

Figure 12. (a) Top- and (b) side-view SEM images of SnO₂ nanorod array. (b)Cycling performances of SnO₂ nanorod array, Disordered SnO₂ nanorods, and SnO₂ nanoparticles. Reproduced from Ref. [53] with permission from The Royal Society of Chemistry.

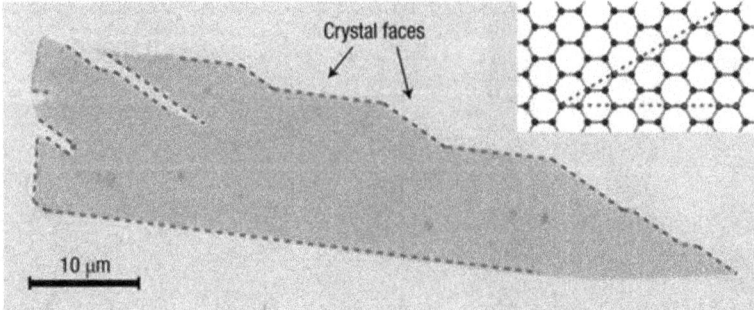

Figure 13. SEM image of a large single layer graphene crystal with crystal's faces demonstrated in inset. Reprinted by permission from Macmillan Publishers Ltd: NATURE MATERIALS ref. [56], copyright 2007.

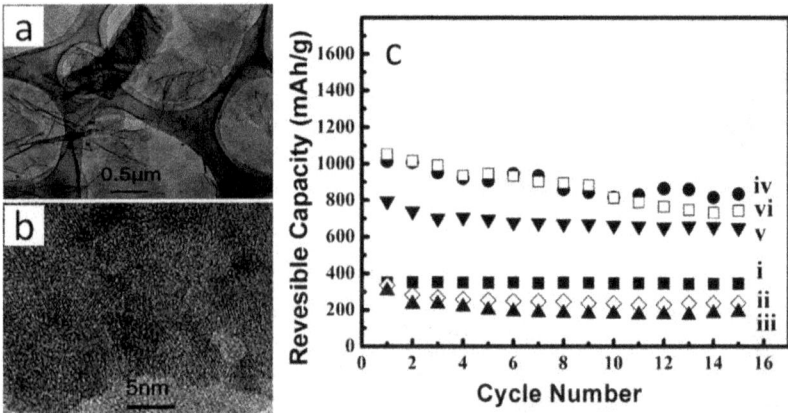

Figure 14. (a) Low magnification and (b) high resolution TEM images of electron-beam-reduced GO. (c) Cycling behaviors of (i)natural graphite, (ii)pristine GO, (iii)hydrazine-reduced GO, (iv)300°C pyrolytic GO, (v)600°C pyrolytic GO, and (vi)electron-beam-reduced GO at a current density of 50 mA g⁻¹. Reprinted with permission from ref. [65]. Copyright 2009 American Chemical Society.

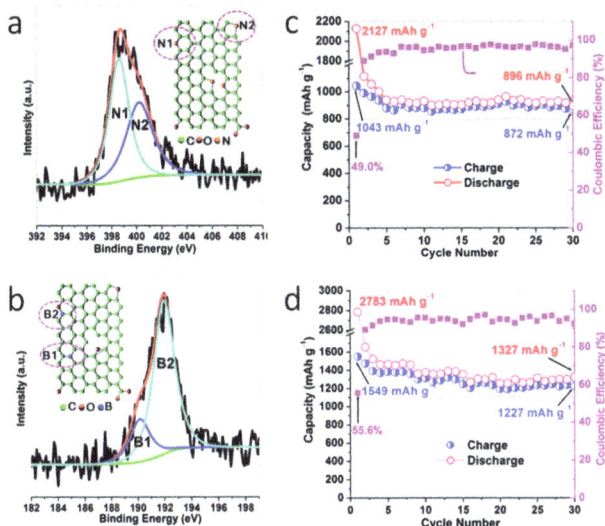

Figure 15. (a) N1s XPS spectrum of N-doped graphene with schematic illustration of the binding condition of N shown in inset. (b) B1s XPS spectrum of B-doped graphene with schematic structure of the binding condition of B shown in inset. Cycling performances of (c) N-doped and (d) B-doped graphene electrodes at a current density of 50 mA g^{-1} between 3 and 0.01 V versus Li^+/Li. Reprinted with permission from ref. [66]. Copyright 2011 American Chemical Society.

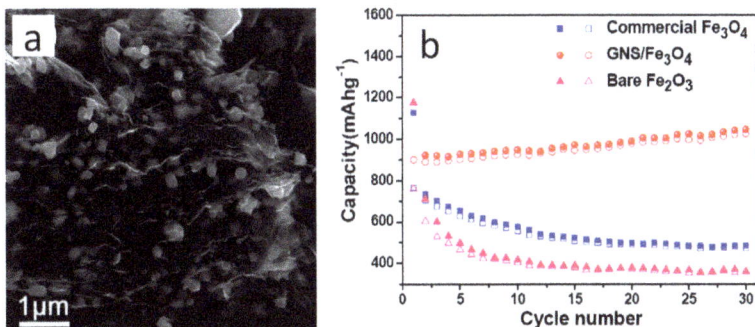

Figure 16. (a) SEM image of GNS/Fe_3O_4 composite. (b) Cycling behaviors of commercial Fe_3O_4, bare Fe_2O_3 particles, and GNS/Fe_3O_4 composite at a current density of 35 mA g^{-1}. Reprinted with permission from ref. [71]. Copyright 2010 American Chemical Society.

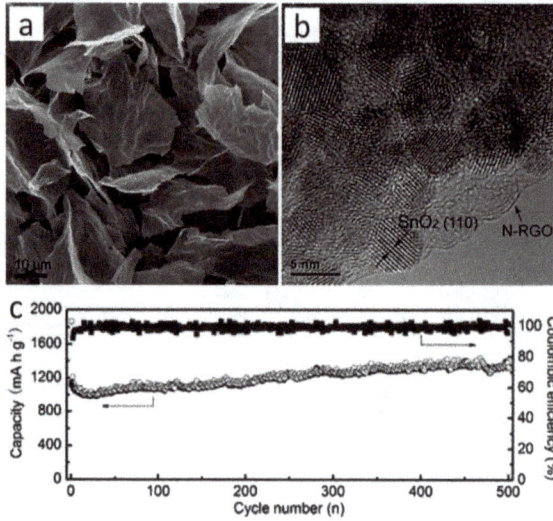

Figure 17. (a) SEM and (b) TEM images of SnO$_2$/N-RGO composite. (c) Cycling performance and Coulombic efficiency of SnO$_2$/N-RGO composite at a current density of 500 mA g^{-1}. Reprinted with permission from ref. [74]. Copyright 2013 Wiley-VCH Verlag GmbH & Co. KGaA, Weinheim.

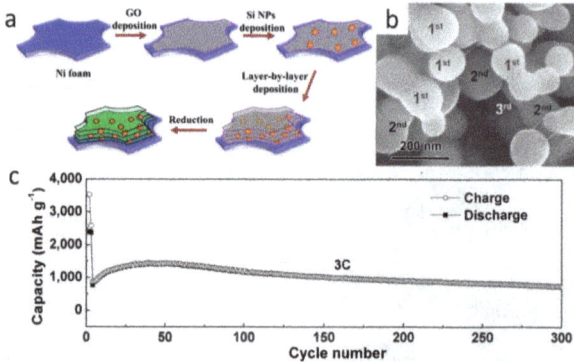

Figure 18. (a) Schematic representation of the fabrication process of Si/RGO composite. (b) SEM image of Si/RGO showing the 1st, 2nd, and 3rd layers of Si nanoparticles. (c) Cycling performance of Si/RGO composite at a rate of 3C for 300 cycles. Reprinted with permission from ref. [86]. Copyright 2013 Wiley-VCH Verlag GmbH & Co. KGaA, Weinheim.

References

[1]   Tarascon, J. M.; Armand, M., Nature 2001, 414 (6861), 359-367.
        https://doi.org/10.1038/35104644

[2]   Goodenough, J. B.; Kim, Y., Chem. Mater. 2009, 22 (3), 587-603.
        https://doi.org/10.1021/cm901452z

[3]   Etacheri, V.; Marom, R.; Elazari, R.; Salitra, G.; Aurbach, D., Energy &
        Environmental Science 2011, 4 (9), 3243-3262.
        https://doi.org/10.1039/c1ee01598b

[4]   Armand, M.; Tarascon, J. M., Nature 2008, 451 (7179), 652-657.
        https://doi.org/10.1038/451652a

[5]   Dylla, A. G.; Henkelman, G.; Stevenson, K. J., Acc. Chem. Res. 2013, 46 (5),
        1104-1112. https://doi.org/10.1021/ar300176y

[6]   Obrovac, M. N.; Chevrier, V. L., Chem. Rev. 2014, 114 (23), 11444-11502.
        https://doi.org/10.1021/cr500207g

[7]   Cabana, J.; Monconduit, L.; Larcher, D.; Palacín, M. R., Adv. Mater. 2010, 22
        (35), E170-E192. https://doi.org/10.1002/adma.201000717

[8]   Kim, T.-H.; Park, J.-S.; Chang, S. K.; Choi, S.; Ryu, J. H.; Song, H.-K., Advanced
        Energy Materials 2012, 2 (7), 860-872. https://doi.org/10.1002/aenm.201200028

[9]   Bruce, P. G.; Scrosati, B.; Tarascon, J.-M., Angew. Chem. Int. Ed. 2008, 47 (16),
        2930-2946. https://doi.org/10.1002/anie.200702505

[10]   Poizot, P.; Laruelle, S.; Grugeon, S.; Dupont, L.; Tarascon, J. M., Nature 2000,
        407, 496-499. https://doi.org/10.1038/35035045

[11]   Guo, Y.-G.; Hu, J.-S.; Wan, L.-J., Adv. Mater. 2008, 20 (15), 2878-2887.
        https://doi.org/10.1002/adma.200800627

[12]   Sim, S.; Cho, J., J. Electrochem. Soc. 2012, 159 (5), A669-A672.
        https://doi.org/10.1149/2.081205jes

[13]   Liang, W.; Yang, H.; Fan, F.; Liu, Y.; Liu, X. H.; Huang, J. Y.; Zhu, T.; Zhang, S.,
        ACS Nano 2013, 7 (4), 3427-33. https://doi.org/10.1021/nn400330h

[14]   Zhang, W.-M.; Wu, X.-L.; Hu, J.-S.; Guo, Y.-G.; Wan, L.-J., Adv. Funct. Mater.
        2008, 18 (24), 3941-3946. https://doi.org/10.1002/adfm.200801386

[15]   Su, L.; Zhou, Z.; Qin, X.; Tang, Q.; Wu, D.; Shen, P., Nano Energy 2013, 2 (2),
        276-282. https://doi.org/10.1016/j.nanoen.2012.09.012

[16] Narayanaswamy, A.; Xu, H.; Pradhan, N.; Peng, X., Angew. Chem. Int. Ed. 2006, 45 (32), 5361-5364. https://doi.org/10.1002/anie.200601553

[17] Liu, S.; Lu, X.; Xie, J.; Cao, G.; Zhu, T.; Zhao, X., ACS applied materials & interfaces 2013, 5 (5), 1588-1595. https://doi.org/10.1021/am302124f

[18] Wang, Z.; Wang, Z.; Liu, W.; Xiao, W.; Lou, X. W., Energy & Environmental Science 2013, 6 (1), 87-91. https://doi.org/10.1039/C2EE23330D

[19] He, M.; Yuan, L.; Hu, X.; Zhang, W.; Shu, J.; Huang, Y., Nanoscale 2013, 5 (8), 3298-3305. https://doi.org/10.1039/c3nr34133j

[20] Lou, X. W.; Li, C. M.; Archer, L. A., Adv. Mater. 2009, 21 (24), 2536-2539. https://doi.org/10.1002/adma.200803439

[21] Chen, Y.; Huang, Q. Z.; Wang, J.; Wang, Q.; Xue, J. M., J. Mater. Chem. 2011, 21 (43), 17448-17453. https://doi.org/10.1039/c1jm13572d

[22] Park, Y.; Choi, N.-S.; Park, S.; Woo, S. H.; Sim, S.; Jang, B. Y.; Oh, S. M.; Park, S.; Cho, J.; Lee, K. T., Advanced Energy Materials 2013, 3 (2), 206-212. https://doi.org/10.1002/aenm.201200389

[23] Zhang, G.; Yu, L.; Wu, H. B.; Hoster, H. E.; Lou, X. W., Adv Mater 2012, 24 (34), 4609-13. https://doi.org/10.1002/adma.201201779

[24] Xia, Y.; Xiao, Z.; Dou, X.; Huang, H.; Lu, X.; Yan, R.; Gan, Y.; Zhu, W.; Tu, J.; Zhang, W.; Tao, X., ACS Nano 2013, 7 (8), 7083-92. https://doi.org/10.1021/nn4023894

[25] Wang, Y.; Cai, L.; Xia, Y., Adv. Mater. 2005, 17 (4), 473-477. https://doi.org/10.1002/adma.200401416

[26] Yin, Y.; Erdonmez, C. K.; Cabot, A.; Hughes, S.; Alivisatos, A. P., Adv. Funct. Mater. 2006, 16 (11), 1389-1399. https://doi.org/10.1002/adfm.200600256

[27] Peng, S.; Sun, S., Angew. Chem. Int. Ed. 2007, 46 (22), 4155-4158. https://doi.org/10.1002/anie.200700677

[28] Son, Y.; Son, Y.; Choi, M.; Ko, M.; Chae, S.; Park, N.; Cho, J., Nano Lett. 2015, 15 (10), 6914-6918. https://doi.org/10.1021/acs.nanolett.5b02842

[29] Liang, J.; Han, X.; Li, Y.; Ye, K.; Hou, C.; Yu, K., New J. Chem. 2015, 39 (4), 3145-3149. https://doi.org/10.1039/C4NJ02313G

[30] Park, M.-H.; Cho, Y.; Kim, K.; Kim, J.; Liu, M.; Cho, J., Angew. Chem. Int. Ed. 2011, 50 (41), 9647-9650. https://doi.org/10.1002/anie.201103062

[31]   Teo, J. J.; Chang, Y.; Zeng, H. C., Langmuir 2006, 22 (17), 7369-7377.
       https://doi.org/10.1021/la060439q

[32]   Liu, S.; Xing, R.; Lu, F.; Rana, R. K.; Zhu, J.-J., The Journal of Physical
       Chemistry C 2009, 113 (50), 21042-21047. https://doi.org/10.1021/jp907296n

[33]   Yang, H. G.; Zeng, H. C., The Journal of Physical Chemistry B 2004, 108 (11),
       3492-3495. https://doi.org/10.1021/jp0377782

[34]   Yang, H. G.; Zeng, H. C., Angew. Chem. Int. Ed. 2004, 43 (44), 5930-5933.
       https://doi.org/10.1002/anie.200461129

[35]   Cao, L.; Chen, D.; Caruso, R. A., Angew. Chem. Int. Ed. Engl. 2013, 52 (42),
       10986-91. https://doi.org/10.1002/anie.201305819

[36]   Chen, Y.; Xia, H.; Lu, L.; Xue, J., J. Mater. Chem. 2012, 22 (11), 5006-5012.
       https://doi.org/10.1039/c2jm15440d

[37]   Sun, Y.; Feng, X.-Y.; Chen, C.-H., J. Power Sources 2011, 196 (2), 784-787.
       https://doi.org/10.1016/j.jpowsour.2010.07.065

[38]   Chen, Y.; Song, B.; Tang, X.; Lu, L.; Xue, J., J. Mater. Chem. 2012, 22 (34),
       17656-17662. https://doi.org/10.1039/c2jm32057f

[39]   Fang, Y.; Gu, D.; Zou, Y.; Wu, Z.; Li, F.; Che, R.; Deng, Y.; Tu, B.; Zhao, D.,
       Angew. Chem. Int. Ed. 2010, 49 (43), 7987-7991.
       https://doi.org/10.1002/anie.201002849

[40]   Chen, Y.; Song, B.; Li, M.; Lu, L.; Xue, J., Adv. Funct. Mater. 2014, 24 (3), 319-
       326. https://doi.org/10.1002/adfm.201300872

[41]   Moon, H. R.; Lim, D. W.; Suh, M. P., Chem Soc Rev 2013, 42 (4), 1807-24.
       https://doi.org/10.1039/C2CS35320B

[42]   Xuan, W.; Zhu, C.; Liu, Y.; Cui, Y., Chem Soc Rev 2012, 41 (5), 1677-1695.
       https://doi.org/10.1039/C1CS15196G

[43]   Zhang, W.; Liu, Y.; Lu, G.; Wang, Y.; Li, S.; Cui, C.; Wu, J.; Xu, Z.; Tian, D.;
       Huang, W.; DuCheneu, J. S.; Wei, W. D.; Chen, H.; Yang, Y.; Huo, F., Adv Mater
       2015, 27 (18), 2923-2929. https://doi.org/10.1002/adma.201405752

[44]   Wu, R.; Qian, X.; Yu, F.; Liu, H.; Zhou, K.; Wei, J.; Huang, Y., Journal of
       Materials Chemistry A 2013, 1 (37), 11126-11129.
       https://doi.org/10.1039/c3ta12621h

[45]   Mai, L.; Tian, X.; Xu, X.; Chang, L.; Xu, L., Chem. Rev. 2014, 114 (23), 11828-
       11862. https://doi.org/10.1021/cr500177a

[46]   Liu, J.; Song, K.; Zhu, C.; Chen, C. C.; van Aken, P. A.; Maier, J.; Yu, Y., ACS
        Nano 2014, 8 (7), 7051-7059. https://doi.org/10.1021/nn501945f

[47]   Chen, Y.; Liu, L.; Xiong, J.; Yang, T.; Qin, Y.; Yan, C., Adv. Funct. Mater. 2015,
        25 (43), 6701-6709. https://doi.org/10.1002/adfm.201503206

[48]   Lu, Y.; Tu, J.; Xiong, Q.; Qiao, Y.; Zhang, J.; Gu, C.; Wang, X.; Mao, S. X.,
        Chemistry – A European Journal 2012, 18 (19), 6031-6038.
        https://doi.org/10.1002/chem.201103724

[49]   Yan, C.; Chen, G.; Zhou, X.; Sun, J.; Lv, C., Adv. Funct. Mater. 2016, 26 (9),
        1428-1436. https://doi.org/10.1002/adfm.201504695

[50]   Zhou, W.; Cheng, C.; Liu, J.; Tay, Y. Y.; Jiang, J.; Jia, X.; Zhang, J.; Gong, H.;
        Hng, H. H.; Yu, T.; Fan, H. J., Adv. Funct. Mater. 2011, 21 (13), 2439-2445.
        https://doi.org/10.1002/adfm.201100088

[51]   Gu, X.; Chen, L.; Ju, Z. C.; Xu, H. Y.; Yang, J.; Qian, Y. T., Adv. Funct. Mater.
        2013, 23 (32), 4049-4056. https://doi.org/10.1002/adfm.201203779

[52]   Yu, L.; Wang, Z.; Zhang, L.; Wu, H. B.; Lou, X. W., Journal of Materials
        Chemistry A 2013, 1 (1), 122-127. https://doi.org/10.1039/C2TA00223J

[53]   Liu, J.; Li, Y.; Huang, X.; Ding, R.; Hu, Y.; Jiang, J.; Liao, L., J. Mater. Chem.
        2009, 19 (13), 1859-1864. https://doi.org/10.1039/b817036c

[54]   Liao, J. Y.; Manthiram, A., Advanced Energy Materials 2014, 4 (14), n/a-n/a.

[55]   Cao, F.-F.; Deng, J.-W.; Xin, S.; Ji, H.-X.; Schmidt, O. G.; Wan, L.-J.; Guo, Y.-
        G., Adv. Mater. 2011, 23 (38), 4415-4420.
        https://doi.org/10.1002/adma.201102062

[56]   Geim, A. K.; Novoselov, K. S., Nature materials 2007, 6 (3), 183-191.
        https://doi.org/10.1038/nmat1849

[57]   Li, X.; Cai, W.; Colombo, L.; Ruoff, R. S., Nano Lett. 2009, 9 (12), 4268-4272.
        https://doi.org/10.1021/nl902515k

[58]   Emtsev, K. V.; Bostwick, A.; Horn, K.; Jobst, J.; Kellogg, G. L.; Ley, L.;
        McChesney, J. L.; Ohta, T.; Reshanov, S. A.; Rohrl, J.; Rotenberg, E.; Schmid, A.
        K.; Waldmann, D.; Weber, H. B.; Seyller, T., Nature materials 2009, 8 (3), 203-
        207. https://doi.org/10.1038/nmat2382

[59]   Wang, J.; Manga, K. K.; Bao, Q.; Loh, K. P., J. Am. Chem. Soc. 2011, 133 (23),
        8888-8891. https://doi.org/10.1021/ja203725d

[60] Brodie, B. C., Philosophical Transactions of the Royal Society of London 1859, 149, 249-259. https://doi.org/10.1098/rstl.1859.0013

[61] Hummers, W. S.; Offeman, R. E., J. Am. Chem. Soc. 1958, 80 (6), 1339-1339. https://doi.org/10.1021/ja01539a017

[62] Dimiev, A. M.; Tour, J. M., ACS Nano 2014, 8 (3), 3060-3068. https://doi.org/10.1021/nn500606a

[63] Dreyer, D. R.; Park, S.; Bielawski, C. W.; Ruoff, R. S., Chem Soc Rev 2010, 39 (1), 228-240. https://doi.org/10.1039/B917103G

[64] Valota, A. T.; Kinloch, I. A.; Novoselov, K. S.; Casiraghi, C.; Eckmann, A.; Hill, E. W.; Dryfe, R. A. W., ACS Nano 2011, 5 (11), 8809-8815. https://doi.org/10.1021/nn202878f

[65] Pan, D.; Wang, S.; Zhao, B.; Wu, M.; Zhang, H.; Wang, Y.; Jiao, Z., Chem. Mater. 2009, 21 (14), 3136-3142. https://doi.org/10.1021/cm900395k

[66] Wu, Z.-S.; Ren, W.; Xu, L.; Li, F.; Cheng, H.-M., ACS Nano 2011, 5 (7), 5463-5471. https://doi.org/10.1021/nn2006249

[67] Zhang, C.; Mahmood, N.; Yin, H.; Liu, F.; Hou, Y., Adv Mater 2013, 25 (35), 4932-4937. https://doi.org/10.1002/adma.201301870

[68] Xu, J.; Lin, Y.; Connell, J. W.; Dai, L., Small 2015, 11 (46), 6179-6185. https://doi.org/10.1002/smll.201501848

[69] Chen, Y.; Song, B.; Tang, X.; Lu, L.; Xue, J., Small 2014, 10 (8), 1536-1543. https://doi.org/10.1002/smll.201302879

[70] Li, N.; Zhou, G.; Li, F.; Wen, L.; Cheng, H.-M., Adv. Funct. Mater. 2013, 23 (43), 5429-5435. https://doi.org/10.1002/adfm.201300495

[71] Zhou, G.; Wang, D.-W.; Li, F.; Zhang, L.; Li, N.; Wu, Z.-S.; Wen, L.; Lu, G. Q.; Cheng, H.-M., Chem. Mater. 2010, 22 (18), 5306-5313. https://doi.org/10.1021/cm101532x

[72] Chen, Y.; Song, B.; Lu, L.; Xue, J., Nanoscale 2013, 5 (15), 6797-6803. https://doi.org/10.1039/c3nr01826a

[73] Chen, Y.; Song, B.; Chen, R. M.; Lu, L.; Xue, J., Journal of Materials Chemistry A 2014, 2 (16), 5688-5695. https://doi.org/10.1039/c3ta14745b

[74] Zhou, X.; Wan, L. J.; Guo, Y. G., Adv Mater 2013, 25 (15), 2152-2157. https://doi.org/10.1002/adma.201300071

[75]   Wu, Z.-S.; Ren, W.; Wen, L.; Gao, L.; Zhao, J.; Chen, Z.; Zhou, G.; Li, F.; Cheng,
        H.-M., ACS Nano 2010, 4 (6), 3187-3194. https://doi.org/10.1021/nn100740x

[76]   Huang, X.-l.; Wang, R.-z.; Xu, D.; Wang, Z.-l.; Wang, H.-g.; Xu, J.-j.; Wu, Z.;
        Liu, Q.-c.; Zhang, Y.; Zhang, X.-b., Adv. Funct. Mater. 2013, 23 (35), 4345-4353.
        https://doi.org/10.1002/adfm.201203777

[77]   Li, L.; Raji, A. R.; Tour, J. M., Adv Mater 2013, 25 (43), 6298-6302.
        https://doi.org/10.1002/adma.201302915

[78]   Wang, H.; Cui, L.-F.; Yang, Y.; Sanchez Casalongue, H.; Robinson, J. T.; Liang,
        Y.; Cui, Y.; Dai, H., J. Am. Chem. Soc. 2010, 132 (40), 13978-13980.
        https://doi.org/10.1021/ja105296a

[79]   Etacheri, V.; Yourey, J. E.; Bartlett, B. M., ACS Nano 2014, 8 (2), 1491-1499.
        https://doi.org/10.1021/nn405534r

[80]   Chang, K.; Geng, D.; Li, X.; Yang, J.; Tang, Y.; Cai, M.; Li, R.; Sun, X.,
        Advanced Energy Materials 2013, 3 (7), 839-844.
        https://doi.org/10.1002/aenm.201201108

[81]   Jiang, Z.; Wang, C.; Du, G.; Zhong, Y. J.; Jiang, J. Z., J. Mater. Chem. 2012, 22
        (19), 9494-9496. https://doi.org/10.1039/c2jm30856h

[82]   Qin, J.; He, C.; Zhao, N.; Wang, Z.; Shi, C.; Liu, E.-Z.; Li, J., ACS Nano 2014.

[83]   Zhao, X.; Hayner, C. M.; Kung, M. C.; Kung, H. H., Advanced Energy Materials
        2011, 1 (6), 1079-1084. https://doi.org/10.1002/aenm.201100426

[84]   Chou, S.-L.; Wang, J.-Z.; Choucair, M.; Liu, H.-K.; Stride, J. A.; Dou, S.-X.,
        Electrochem. Commun. 2010, 12 (2), 303-306.
        https://doi.org/10.1016/j.elecom.2009.12.024

[85]   Zhang, J.; Zhang, L.; Xue, P.; Zhang, L. Y.; Zhang, X. L.; Hao, W. W.; Tian, J.
        H.; Shen, M.; Zheng, H. H., Journal of Materials Chemistry A 2015, 3 (15), 7810-
        7821. https://doi.org/10.1039/C5TA00457H

[86]   Chang, J.; Huang, X.; Zhou, G.; Cui, S.; Hallac, P. B.; Jiang, J.; Hurley, P. T.;
        Chen, J., Adv Mater 2014, 26 (5), 758-764.
        https://doi.org/10.1002/adma.201302757

CHAPTER 6

Electrolytes for Advanced LIBs

Jinkui Feng

School of Materials Science and Engineering, Shandong University, jinkui@sdu.edu.cn

Abstract

Electrolytes are essential parts for LIBs. The quality of electrolytes restricts the performance of the integrated LIBs. The developments of LIBs demand safer, wider working potentials and better compatible electrolytes to satisfying the emerging high voltage cathodes, such as $LiNi_{0.5}Mn_{1.5}O_4$, Lithium-rich Mn-based layered cathode and $LiCoPO_4$ etc., high capacity anodes such as Si, Sn and lithium metal etc. and the safety concerns such as burning and explosive. This chapter reviews the recent progress in electrolytes used to improve performance and other properties of LIBs, such as safety. This chapter classified the electrolytes based on the different compositions of electrolytes: solvents, lithium salts and additives.

Keywords

Electrolytes, Solvents, Lithium Salts, Safe, Additives

Contents

1.   Introduction

Electrochemistry studies chemical reactions which take place at the interface of electronic conductive electrodes and an ionic conductor - the electrolyte. Electrolytes are essential to all kinds of electrochemical devices, the function of electrolytes in batteries serve as the medium for the transfer of ions between the electrodes. The general electrolytes consist of salts dissolved in solvents. Because of its physical location in the electrochemical devices, that is, bridging positive and negative electrodes, the electrolyte is in close interaction with both electrodes. The interfaces between the electrolyte and the two electrodes often decide the performance of the electrochemical devices. In fact, these electrified interfaces have been the focus of interest since the development of electrochemistry.

In a battery, the electrolyte defines how fast the energy can be released by controlling the rate of ion transfer between the cathodes and the anodes. The ideal electrolyte should undergo no chemical changes during the operation of the battery. It must be stable against both cathodes and anodes. A generalized list of these minimal requirements should include the following [1]:

1) It should be a good ionic conductor and electronic insulator, so that ion transport can be facile and self-discharge can be kept to a minimum.

2) It should have a wide electrochemical window, so that electrolyte degradation would not occur within the range of the working potentials of both the cathode and the anode.

3) It should also be inert to other cell components such as cell separators, electrode substrates, and cell packaging materials.

4) It should be robust against various abuses, such as electrical, mechanical, or thermal ones.

5) Its components should be environmentally.

Lithium has long received much attention as a promising anode material. The interest in this alkali metal has arisen from the combination of its two unique properties: (1) it is the most electronegative metal (-3.045 V vs. SHE), and (2) it is the lightest metal (0.534 g cm^{-3}). The former confers upon it a negative potential that translates into high cell voltage when matched with certain cathodes, and the latter makes it an anode of high specific capacity (3860 mAh g^{-1}). In the 1950s, lithium metal was found to be stable in a number of nonaqueous solvents such as propylene carbonate (PC) and tetrahydrofuran (THF), and this stabilization was attributed to the formation of a passivation film on the lithium metal surface, which prevents it from continuous reaction with electrolytes. So the concept of lithium batteries is proposed. A series of lithium metal based primary cells come into market from 1960s, and the electrolyte solvents ranged from organic (propylene carbonate) to inorganic (thionyl chloride or sulfur dioxide). However, dendrite formation problems failed the commercialization of lithium metal based secondary batteries. In 1989, incidents of fire due to lithium rechargeable batteries in the electronic devices excluded lithium metal based secondary batteries [1].

In 1990s, with carbonaceous substitute for lithium metal, lithium ion battery comes into commercialization. The advantage of carbonaceous anode is the low cost, the high safe character and the low lithium intercalation plateau. In 1990 both Sony and Moli announced the commercialization of lithium ion batteries based on petroleum coke and $LiCoO_2$ with PC based electrolyte. However, PC is not stable with the high energy graphitic base anodes. Dahn et al published their pioneering work on the electrolytes for graphitic anodes and the effect of electrolyte solvent. They found that ethylene carbonate (EC) is stable with graphitic anode and propose the interface between solvents and anodes [2]: Electrolyte solvents decompose reductively on the carbonaceous anode, and the decomposition product forms a protective film, called "solid electrolyte interface" SEI films. The film is ionic conductive and electron insulating which could prevents further decomposition of the electrolyte components. The SEI film formed on a carbonaceous anode plays a critical role in enabling lithium ion batteries to work reversibly. SEI film consists of thermodynamically stable salt such as Li_2O, LiF, Li_2CO_3 et al located in the inner surface of the anode and organic species such as $ROCO_2Li$, $ROLi$, $(ROCO_2Li)_2$ and polycarbonates located closer to the electrolyte solution. The SEI film of carbonaceous model is the same as proposed earlier by Peled in 1979 to describe the passivation on lithium metal [3-7].

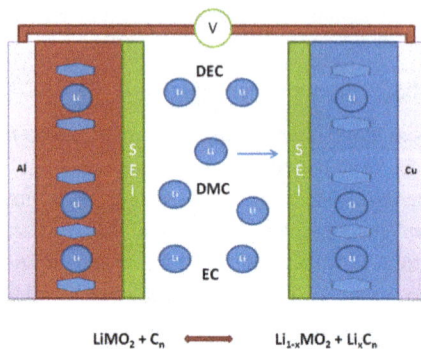

Figure 1. Schematic illustration of lithium ion battery

However, the property of single solvent cannot perfectly suit for the demands of lithium ion batteries. For example, high dielectric constant always comes with high viscosity. Most lithium electrolytes are composited of one lithium salt with two or more solvents. The state-of-art electrolytes are composited of cyclic and linear carbonates solvents, lithium salts (mainly $LiPF_6$) and additives to improve the performance. The electrolytes satisfy the basic demands of ongoing 4 V lithium ion batteries. However, Lucht et al found that 1 M $LiPF_6$ in ethylene carbonate/dimethyl carbonate/diethyl carbonate (1:1:1) could electropolymerization on the $LiNi_{0.5}Mn_{1.5}O_4$ surface. XPS proved that EC starts to polymerization from 4.5V (vs. Li/Li^+) [8].

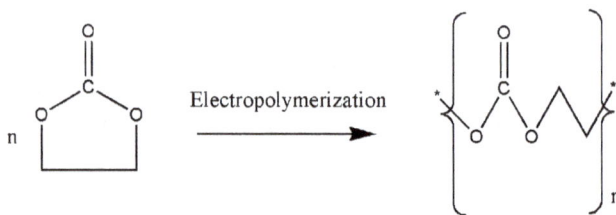

Figure 2, the schematic of electropolymerization of EC

Moreover, the carbonates are highly flammable [9]. The developing of high energy lithium ion batteries and the demanding of extended usage, such as 5V class cathode, silicon anode, Electric Vehicles and large energy storage equipment, require advanced electrolyte systems. The requirements can be summarized as below:

1) Prolonged cycling life. As the warrant time of vehicles could be higher than 5 years, cycling performance of lithium batteries should be as long as possible.

2) Safer. The existing lithium ion batteries make use of highly flammable electrolytes and highly reactive electrode, when fall under abused conditions, the batteries may be explosive and burning.

3) Fits new materials, such as high voltage cathodes, lithium metal anode and alloying anode.

4) More tolerance to extreme conditions, such as lower and higher temperature, high rate et al.

2. Solvents

In accordance with the basic requirements for electrolytes, an ideal electrolyte solvent should meet the following minimal demand [1]:

1) It should be able to dissolve salts to sufficient concentration. In other words, it should have a high dielectric constant.

2) It should be fluid (low viscosity η), so that facile ion transport can occur.

3) It should remain inert to all cell components, especially the charged surfaces of the cathode and the anode, during cell operation.

4) It should remain liquid in a wide temperature range. In other words, its melting point should be low and its boiling point high.

5) It should also be safe, nontoxic, and economical.

For lithium-based batteries, the active nature of the strongly reducing anodes (lithium metal or the highly lithiated carbon) and the strongly oxidizing cathodes (transition metal based oxides) rules out the use of any solvents that have active protons despite their excellent power in solvating salts, because the reduction of such protons and/or the oxidation of the corresponding anions generally occurs within 2.0- 4.0 V versus Li, while the charged potentials of the anode and the cathode in the current rechargeable lithium devices average 0.0-0.2 V and 3.0-4.5 V, respectively. On the other hand, the nonaqueous compounds that qualify as electrolyte solvents must be able to dissolve sufficient amounts of lithium salt; therefore, only those with polar groups such as carbonyl, nitrile, sulfonyl, and ethers et al [1].

Among them, only carbonate based electrolytes could form stable SEI film on the carbon based anode materials. While all of the ethers, cyclic or linear, demonstrate similar moderate dielectric constants (2-7) and low viscosities (0.3-0.6 cP), cyclic andacyclic esters behave like two entirely different kinds of compounds in terms of dielectric constant and viscosity; that is, all cyclic esters are uniformly polar (40-90) and rather viscous (1.7-2.0 cP), and all acyclic esters are weakly polar (3-6) and fluid (0.4-0.7 cP). The origin for the effect of molecular cyclicity on the dielectric constant has been attributed to the intramolecular strain of the cyclic structures that favors the conformation of better alignment of molecular dipoles, while the more flexible and open structure of linear carbonates results in the mutual cancellation of these dipoles.

EC has comparable viscosity and high dielectric constant (89.8), which are favorable for a solvent candidate. However, its high melting point (~36 °C) and high viscosity limits its working range. By introducing linear carbonate electrolyte with low viscosity and low melting point, we can get a combined electrolyte with wide potential window, wide temperature range and fast ion transport. The hybrid solvent system is now successfully applied in the commercial lithium ion battery. However, with the developing of electrode materials and wider application aspect, the limitations of the state-of-art solvent system arise. Such as highly flammable, easy to be oxidized when charged above 4.3V AND poor SEI film stability et al [8].

2.1 The fluoro-substituted solvents

The carbonate solvents are most flammable and voltage sensitive, when subscribed to abused conditions such as heating, ignition, pouch, crash or overcharged, it can easily react with the highly reactive cathode and anode, results in firing or explosive. Numerous incidents have been reported. To overcome the safety problem, new safe solvents are needed. Due to the catalyst character of transition metal ions, most carbonate electrolyte will decompose at 4.5V (vs. Li/Li$^+$) [8]. With the developing of novel high voltage electrode materials, such as $LiNi_{0.5}Mn_{1.5}O_4$, Mn-rich cathode et al, anodic oxidation resistance solvents are demanding. Research proved that electrolyte oxidation is the major problem that causes cell failure for NMC/graphite cells charged to high potentials. Great effort has been put on developing new solvent for advanced lithium ion batteries.

To overcome the anodic limitation of carbonate solvent, great effort has been put on developing high voltage solvents. It is proved that the calculated energies for the lowest unoccupied molecular orbital (LUMO) can be correlated with oxidation potentials, lower LUMO levels result in higher oxidation stability. Research indicated that fluorinated electrolytes have lower LUMO and are more stable than the non-fluorinated counterparts. For example, K. Amine found that the oxidation potential of Fluorinated EMC can be greatly increased from 6.63 to 7.01 V (vs. Li/Li$^+$), By comparing the conventional Gen 2 electrolyte (1.2 M LiPF$_6$ dissolved in ethylene carbonate (EC) and ethyl methyl carbonate (EMC) in 3:7 ratio by weight) and the Fluorinated HVE electrolyte (1.0 M LiPF$_6$ in

Fluoroethylene carbonate (FEC), Methyl 2,2,2-trifluoroethyl carbonate (F-EMC), 1,1,2,2-tetrafluoroethyl-2,2,3,3-tetrafluoropropyl ether (F-EPE) at 3/5/2 ratio by volume), the HVE showed much improved cycling performance and coulombic efficiency at room temperature and 55 degree. Long cycling data of LNMO/graphite cells show that the HVE cell is able to retain 50% of its initial capacity over 600 cycles at RT and 250 cycles at 55 °C. These discoveries indicate that the electrolyte plays the most significant role in the high voltage cell performance [10].

Moreover, it is found that Fluorinated solvents can form more stable SEI films on the anode due to its lower HOMO, which results in higher reduction potentials. For example, it is well known that PC is a good candidate for lithium ion batteries due to its wide potential window, higher liquid temperature range and high dielectric constant. However, it is not stable with graphite anode as it cannot form stable SEI films on the graphite anode. By substituting H in PC with F elements, trifluoropropylene carbonate (TFPC) could form a stable SEI film on the graphite anode due to is higher HOMO level. In 1999, Ogumi's group found that lithium could be intercalated into graphite in 1 M $LiClO_4$ trifluoropropylene carbonate (TFPC) as solid electrolyte interface (SEI) film can be formed [11]. Arai et al further found that The ClEC/TFPC electrolyte showed higher discharge capacities with lower irreversible capacity loss in both a graphite/Li cell and $Li_{1+x}Mn_2O_4$/Li cell than non-fluoro-substituted carbonate systems [12]. Zheng et al further proved that TFPC can not only establish a stable SEI layer on graphite electrode and suppress the intercalation reaction of PC molecules, but also the $LiNi_{0.5}Mn_{1.5}O_4$ cathodes were found to exhibit high reversible capacity and superb rate performance in the optimized electrolyte [13]. Combined with the inherent characters of high conductivity, excellent solubility with lithium salts fluorinated cyclic and linear carbonate compounds possess promising properties as solvents for lithium ion batteries. Additionally, fluorinated solvents have lower flammability due to the flame retarding ability of F element and could improve the safety characteristics performance of Li-ion cells [14].

However, fluorinated electrolytes have their internal shortcomings. Since LUMO energy levels are also lowered by fluorinated substitution, the fluorinated electrolytes have higher reduction potential, resulting in instability on the anode side of the high-voltage graphite/LNMO cells due to the higher reduction potential. If the fluorinated solvents in the electrolyte decompose prior to this potential of EC and cannot form an effective SEI, they will cause continuous decomposition leading to the failure of the battery. Fluorinated ether F-EPE does not possess such a threat. However, when EC is partially or fully replaced by the fluorinated cyclic carbonate, a low initial discharge capacity or low capacity retention was observed due to the extensive decomposition with tremendous lithium loss before the SEI formation that prevents or slows down the decomposition. An efficient SEI additive or high voltage anode such as $Li_4Ti_5O_{12}$ is necessary for the fluorinated electrolytes to stabilize the graphite anode and electrolyte interphase [15].

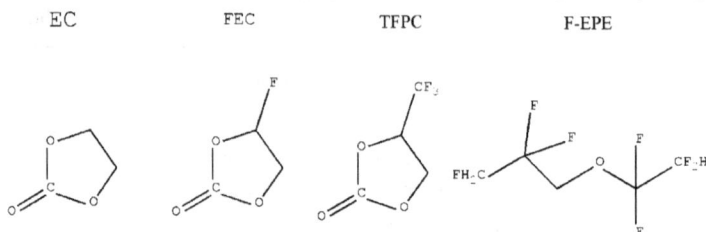

Figure 3, the schematic of structure of EC, FEC, TFPC and F-EPE

Recently, Dahn et al found that NMC442/graphite cells containing fluoroethylene carbonate: bis (2,2,2-trifluoroethyl) carbonate (FEC:TFEC) (1:1 w:w) electrolyte produced large amounts of gas during storage. An SEI film additive -Prop-1-ene-1,3-sultone (PES) addition could reduce the gas amount. The long-term cycling results showed that cells containing PES:FEC:TFEC electrolyte had better capacity retention than cells containing binary or ternary electrolyte additives in EC:EMC solvent. Symmetric cell studies and EIS results suggested that the positive electrode in PES:FEC:TFEC electrolyte is relatively stable at potentials as high as 4.5 V. Nevertheless, the PES:FEC:TFEC electrolyte system has problems. Cells containing PES:FEC:TFEC electrolyte show very large negative electrode impedance which might limit high rate and low temperature applications. Gas evolution during long-term cycling and during charge/hold protocols is large even when gas reducing reagent, PES, is used as an additive [16]. Further work is required to limit negative electrode impedance and limit gas generation.

2.2   Phosphate ester solvents

Safety is a problem of prime concern for development of large size lithium batteries, which are greatly needed in many applications such as electric tools and electric vehicles. Although a number of factors can induce the unsafe behaviors of lithium batteries, one of major sources for safety hazards comes from the oxidation reactions of electrolyte solvent on charged cathodes, which release excessive amount of heat and cause thermal runaway of the cells. As a consequence, the cells may appear to vent and burn due to the ignition of flammable electrolyte leaked at high temperature. Thus, it is expected that the firing and burning of electrolyte solution could be avoided if the electrolyte are non-flammable or fire retardant. Organic phosphates are known as commonly used fire-retardant liquids. Compared with alkyl carbonates, some of these liquid compounds have similar liquidus temperature range, solvating ability and electrochemical stability, seeming to be a good candidate as non-flammable solvents for lithium batteries.

DMMP   **TMP**

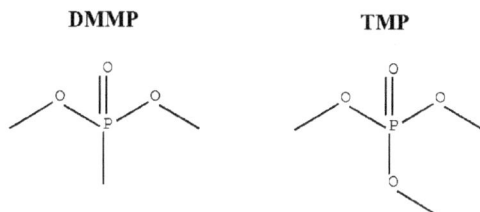

Figure 4, the schematic of structure of DMMP and TMP

DMMP is a polar liquid compound of low viscosity (cP ~1.75, 25 °C), strong solvating ability and wide liquidus temperature range (-50 to 181 °C). Particularly, this compound is incombustible and is now used as a strong fire-retardant in many aspects of fire protections (Table 1).

Table 1 The physical properties comparison of DMMP and PC

Name	m.p/ °C	b.p/ °C	Viscosity	Flash point °C	P %	d/g.cm^{-3}
PC	-48	242	2.53	132	0	1.21
DMMP	-50	181	1.75	104	25	1.16

For most lithium batteries, the electrolyte solvents should have at least a stable potential window of 0-4.5 V (vs. Li) to avoid the electrochemical interference of the electrolyte in the normal charge-discharge of lithium storage anodes and oxide cathodes. The cyclic voltammograms of a number of lithium battery electrode materials in 0.8 mol L^{-1} LiBOB + DMMP electrolyte were probed. As it can be seen, the CV curve from a Pt working electrode in the DMMP electrolyte showed negligible background current in the potential region from 4.5 to 0 V and gave only a pair of lithium deposition and stripping peaks once potential was scanned to -0.5 V. At more positive potentials than 4.5 V, a small oxidation current was observed due to the oxidative decomposition of LiBOB. This suggests that DMMP is very stable in the potential range of 4.5-0 V and is also workable for electrochemical deposition and dissolution of lithium. Similarly, all the cathode materials $LiMn_2O_4$, $LiCoO_2$, $LiFePO_4$ and anode material $Li_4Ti_5O_{12}$ in the DMMP electrolyte exhibited their perfect CV bands, identical to those observed in commercial organic carbonate electrolytes, demonstrating that electrochemical activities of these electrode materials are unaffected in the DMMP electrolyte [17].

Another important property of an electrolyte solvent for lithium batteries is its ability to dissolve sufficient amount of lithium salt. In our experiment, DMMP solvent can dissolve more than 2 mol L^{-1} of LiBOB, $LiPF_6$ or $LiClO_4$ salt, forming considerably conductive electrolyte solutions. It is shown in the figure that the ionic conductivity of the DMMP solution can reach a considerable high value of >3 mS cm^{-1} at a wide concentration range of 0.3-1.2 mol L^{-1} LiBOB electrolyte with its maximum conductivity of 4 mS cm^{-1} at 0.6 mol L^{-1} LiBOB. Though the conductivity is not as high as those reported in organic carbonate electrolytes (6-7 mS cm^{-1}), it is higher than gel electrolyte (1-3 mS cm^{-1}) used in lithium batteries. The conductivity of the DMMP electrolyte varies with temperatures in accord with Arrhenius relation. In comparison, the ionic conductivity of the DMMP-based electrolyte is even higher than the PC-based electrolyte at the same concentration of $LiClO_4$ salt.

The capacity retention of Li-$LiMn_2O_4$ cells cycled in DMMP electrolyte and in TMP electrolyte was examined. The cells using DMMP electrolyte can be cycled very well with only 5% capacity loss during 50 cycles, while the cells using TMP electrolyte showed rapidly declined capacity after a few cycles, implying an electrochemical stability of Li and $LiMn_2O_4$ electrode in the DMMP electrolyte. In previous work, it has been revealed that since the SEI film formed on the cathode surface in TMP electrolyte is not stable, TMP molecules can react continuously with the oxidized cathode to result in continuous degradation of the cathode material. In contrast, DMMP exhibited considerable electrochemical stability at the charged cathodes, and it is therefore expected that the cycleability of Li-$LiMn_2O_4$ cells may be greatly improved by use of DMMP electrolyte.

Though most of the organic phosphates enable to reduce the flammability of electrolyte, they are difficult for practical use because most of the organic phosphates can decompose electrochemically at quite positive potentials, which cannot form a stable solid electrolyte interface (SEI) film so as to prevent the further decomposition of electrolyte. To solve this problem, we tried to use the film-forming additives in the electrolyte to produce a stable SEI film on the graphite surface before the onset of electroreduction of DMMP solvent and thereby to suppress the direct reduction and co-intercalation of DMMP molecules. By comparison of a number of the film-forming molecules, we found that Cl-

EC can function very well for this purpose. The CV of a graphite anode in the DMMP electrolyte containing 10% Cl-EC is probed. At first cathodic scan, a broad irreversible band appeared at ca. +1.7 V, denoting the electrochemical reduction of Cl-EC additive. When the potential scan was swept negatively, the current peaks featuring the decomposition and co-intercalation of DMMP solvent became indiscernible and instead, a pair of redox peaks characteristic of Li^+ interaction and de-intercalation on graphite appeared reversibly at the potential region of +0.5 to 0 V and became a prominent CV feature since the second scan and afterwards. This CV evidence suggested that Cl-EC reduces to form a stable SEI film on the graphite surface prior to the reductive decomposition of DMMP solvent, which protected the graphite structure from destructive attacking by the co-intercalation of the phosphate molecules. The reason why only Cl-EC can effectively form a protecting film on the graphite anode is simply because the reduction potential of Cl-EC ($\sim$1.7 V) is more positive than that of DMMP ($\sim$1.3 V) and the electrochemical reduction of Cl-EC is earlier than the decomposition of DMMP molecules. Also, the reductive decomposition of Cl-EC produce CO_2, which is known to be an important component for promoting the formation of a dense and stable SEI film on graphite surface [18].

The strong ability of Cl-EC additive for improving the electrochemical performance of graphite in the DMMP solution can also be seen from the charge–discharge experiments. The charge-discharge curves of graphite in 1 mol L^{-1} LiClO$_4$ + DMMP with and without addition of Cl-EC were compared. In the absence of Cl-EC, the charging capacity of the graphite anode was very large ($\sim$1200 mAh g−1) and the charging potential showed a plateau at $\sim$1.5 V and then decreased slowly to 0.2 V, suggesting that the charging capacity was mostly consumed in the electrochemical decomposition and co-intercalation of DMMP molecules. Because of this inefficient charging, the graphite electrode gave almost no useful capacity at discharge, showing a rapid rise of the discharging potential. In comparison, the charging potential of the graphite electrode in the presence of Cl-EC additive decreased rapidly from 3 to +0.2 V and then kept stable at +0.2 to +0.05 V, which is characteristic of the formation of LiC_x compound by Li^+ insertion into the graphite. At discharge, the graphite electrode gave a flat discharge plateau at $\sim$0.25 V and delivered a capacity of $\sim$280 mAh g^{-1} corresponding to a capacity utilization of $\geq$85% at the first cycle. This first discharge capacity and coulombic efficiency is very similar to those observed in carbonate electrolytes. From the second cycle, the cycling efficiency of the graphite electrode increased rapidly to $\sim$100% and remained steadily at prolonged cycling. This comparison demonstrates that the graphite electrode can work very well as Li^+ inserting anodes in the DMMP electrolyte as long as a small amount of Cl-EC additive was added to block off the reductive reactions of DMMP molecules [18].

2.3 Sulfone-based solvents

Sulfone solvents are novel kinds of electrolyte for lithium battery due to its high resistance to oxidation and nonflammable character. Sulfone exhibited the highest anodic potential, above 6.0 V versus metallic lithium. Therefore, sulfone-based electrolytes were suitable for use with high voltage electrode such as $LiNi_{0.5}Mn_{1.5}O_4$. In 1998, Angell et al reported the possible using of ethyl methyl sulfone (EMS) as high voltage solvents for lithium ion batteries [19]. However, most sulfones are solid at room temperature. Low melting co-solvents are needed to get a liquidus electrolytes. Angell further found that by introducing a carbonate co-solvent, DMC, a high oxidation limit of >5.9 V vs. Li/Li^+ and an 89.1% initial coulomb efficiency for $LiNi_{0.5}Mn_{1.5}O_4$ cathode could be obtained with 1 M $LiPF_6$ in EMS-DMC (1:1 by wt.) electrolyte. Moreover, capacity retention of $LiNi_{0.5}M_{1.5}O_4$ yields 97% over 100 cycles [20-23]. Another obstacle of the sulfones based solvents is that they cause exfoliation of graphite electrode when used as pure electrolyte in lithium ion cells. This problem can be solved with high voltage anode such as $Li_4Ti_5O_{12}$. It is reported that when $Li_4Ti_5O_{12}$ is used as anode for 1 M $LiPF_6$ in EMS-DMC (1:1 by wt.) electrolyte, the capacity retention is 100% and the coulombic efficiency is remarkably high [23]. Another method to address this problem is introducing SEI formation additives. Lewandowski group found that graphite electrode could deliver a capacity of 350 mAh g^{-1} in the 1 M $LiPF_6$ + TMS + VC 10 wt% electrolyte. VC could act as the SEI film formation additive and prevent the graphite exfoliation. Moreover, The $LiFePO_4$ + 1 M $LiPF_6$ + TMS + 10 wt% VC system shows a flash point of ca. 150 °C, which was much higher in comparison to that characteristic of a classical $LiFePO_4$ + 1 M $LiPF_6$ + 50 wt% EC + 50 wt% DMC system ($T_f \approx 37$ °C) [24]. Mao and Hofmann et al also found that adopting SEI forming lithium salts could also solve the anode compatibility of sulfone-based electrolytes [25-27]. For example. Mao et al found that with Lithium bis(oxalate)borate (LiBOB) as the salt, MCMB (mesophase carbon microbeads) could cycle very well with a capacity retention of 90.91%. Moreover, $LiFePO_4$/Li cells with LiBOB-SL/DMS electrolyte showed excellent cycling ability and good thermal stability at 60 ∘C.

2.4 Ionic liquids

Ionic liquids are salts with low temperature melting points (near room temperature). As novel kind of solvents for lithium batteries, ionic liquids inquire special advantages: 1) Ionic liquids are non-flammable. 2) The thermal stability of ionic liquids is high (higher than 300 ℃) [1,28].With ionic liquids as solvents, the safety character of lithium batteries could be greatly improved. Ionic liquids are composites of large size cations, such as pyridinium, imidazolium, quaternary ammonium, piperidinium, pyrrolidinium, phosphonium et al. The anions are composed of $CF_3SO_3^-$, BF_4^-, PF_6^-, $TFSI^-$ et al [28]. Among these cations, pyridinium and imidazolium cations are not stable with lithium metal, which limits it use in lithium metal batteries. For quaternary ammonium, piperidinium, pyrrolidinium, phosphonium based ionic liquids, only when combined with large ionic radius anions such as TFSI , the melting points lower to room temperature [29-40]. In 2003, Sakaebe et al compared the lithium storage ability of EMI (1-ethyl-3-methylimidazolium)TFSI, TMPA (trimethylpropylammonium)TFSI, P13 (N-methyl-N-propylpyridinium)TFSI, PP13 (N-methyl-N-propylpiperidinium)TFSI as electrolyte for Li/LiCoO$_2$ cell. The cathodic stability of the IL is in the order; EMITFSI < TMPATFSI < P13TFSI or PP13TFSI. Only EMITFSI was positive against Li/Li$^+$. Among these ionic liquids, quaternary ammonium cations could form the stable RTIL against the reduction on the lithium metal. PP13TFSI is the most promising candidate as the electrolyte base with a melting point of 8.7 and a conductivity of 1.51 mS cm^{-1}. Li/LiCoO$_2$ cell containing PP13TFSI showed best performance with a capacity of 120 mAh g^{-1} and the Coulombic efficiency at entire cycles is more than 97%. When cycled at higher rate (C/2), 85% of the discharge capacity was still retained after about 30 cycles. while the cells contain TMPATFSI showed large irreversible capacity and suddenly lost the capacity and P13TFSI showed fast capacity retention.

However, one disadvantage of ionic liquids is that graphite cannot sufficiently operate in many types of ILs. The main reason would be due to the electrochemical cation intercalation into the graphite layers before the formation of an effective solid electrolyte interface (SEI) film on the graphite particles during the 1st lithium intercalation process. To overcome this shortcoming, one effective method is adding SEI film forming additive, such as VC or lithium difluoro(oxalate)borate (LiODFB). Wu found that LiODFB could facilitates the SEI formation on the surfaces of carbonaceous anode. The SEI layers prevent the electrolyte from decomposition and protect the graphite materials from decomposition. The Li/MCMB cell showed a 1st discharge capacity of 369.5 mA h g^{-1} and 338.6 mAh g^{-1} after 50 cycles in the LiODFB-PP$_{14}$TFSI/TMS cell. Sun et al found that 10 wt% of vinylene carbonate (VC) introducing could form stable SEI film in ionic liquids based on methylpropylpyrrolidinium (MPPY) and methylpropylpiperidinium (MPPI) cations and bis(trifluoromethanesulfonyl)imide (TFSI) anion. At 50 ℃ the capacity is increased to 325 mAh g^{-1} for the cell using 0.5 M LiTFSI/MPPY.TFSI +10 wt% VC while it is stabilized at 310 mAh g^{-1} for the cell using 0.5 M LiTFSI/MPPI.TFSI +10 wt% VC. The coulombic efficiencies, except the first two formation cycles, are all above 96% [41].

Adding ether group on the ammonium cation, graphite anode could function well in the N, N-diethyl-N-methyl-N-(2-methoxyethyl) ammonium bis(trifluoromethylsulfonyl) amide (DEME-TFSI) ionic liquid with LiTFSI salt electrolyte. Graphite electrode showed an initial capacity of 318 mAh g^{-1} and initial coulombic efficiency of 75.6%. The discharge capacity maintained at 320 mAh g^{-1} and efficiency were stable at 100% [32].

Recently, it is found that with a new anion bis(fluorosulfonyl)imide (FSI), graphite can cycle very well in the 1-ethyl-3-methylimidazolium bis(fluorosulfonyl)imide (EMImFSI) ionic liquid (EMImFSI)/LiTFSI, the graphite electrodes exhibited a high coulombic efficiency (greater than 80%) that was comparable to their efficiency in $LiPF_6$/EC+DMC electrolyte and a high reversible capacity of 365 mAh g^{-1}. A mechanism involving a double-layer-based interface is proposed for the FSI based ionic liquids [39].

2.5   Water based solvent

The high flammable and low conductivity of organic solvents restrict the wide application of lithium batteries. Water has a large dielectric constant (78.36F/m), low viscosity (about 1mPa/s at 293K), low cost, environmental friendly and nonflammable character, enable it a promising candidate for electrolyte. However, the low potential window and reactivity with lithium metal are the restriction. By choosing the high intercalation voltage anode, water can be used as the solvent for lithium ion batteries. In 1994, Dahn's group first reported the aqueous lithium ion batteries with $LiMn_2O_4$ as cathode and $VO_2(B)$ as anodes and 5 M $LiNO_3$ in water as the electrolyte [42]. The weight energy density of the battery system can compete with nickel-cadmium and lead-acid batteries. However, serious capacity fading hinder its further applications. To improve the cycle ability and increase the energy density, recently, with the developing of novel electrolytes and electrodes, the aqueous rechargeable lithium ion batteries (ARLB) attracted more attentions. New electrodes including $LiTi_2(PO_4)_3$, TiO_2, LiV_3O_8, MoO_3, $LiFePO_4$, $LiMnPO_4$, $Li(NiCoMn)O_2$ et al are developed. New salts such as Li_2SO_4, $LiClO_4$ are probed [43-47]. Research found that the carbon coating, pH value, lithium salt concentration and oxygen content play critical roles to the electrochemical performance of ARLB. By eliminating the oxygen and adjust the pH values in the water, Xia et al found that the $LiTi_2(PO_4)_3$-$LiFePO_4$ battery system with 0.5 M Li_2SO_4 aqueous electrolyte could retain 90% capacity after 1000 cycles [43]. However, the low operating voltage and the high cost of lithium reduced its economic efficiency compared to other aqueous battery systems.

In 2015, a new "water-in-salt" aqueous electrolyte was report by Wang et al. By dissolving high concentrations (>20M) LiTFSI in water, the electrochemical window could be extended to about 3.0 V due to the formation of protective SEI layer between the Mo_6S_8 anode and electrolyte.The SEI layer was formed by the electrolyte decomposition during the initial charging processes. A 2.3 V lithium-ion battery of using this aqueous electrolyte could deliver a 100% coulombic efficiency and a cycle life of more than 1000 cycles. The system could compete with the conventional non-aqueous Li-ion batteries in terms of power and energy density. This electrolyte shines new light on the development of aqueous electrolyte [48].

3. Lithium salts

An ideal electrolyte solute for ambient rechargeable lithium batteries should meet the following minimal requirements: (1) It should be able to completely dissolve and dissociate in the nonaqueous media, and the solvated ions (especially lithium cation) should be able to move in the media with high mobility. (2) The anion should be stable against oxidative decomposition at the cathode. (3) The anion should be inert to electrolyte solvents. (4) Both the anion and the cation should remain inert toward the other cell components such as separator, electrode substrate, and cell packaging materials. (5) The anion should be nontoxic and remain stable against thermally induced reactions with electrolyte solvents and other cell components [1,49].

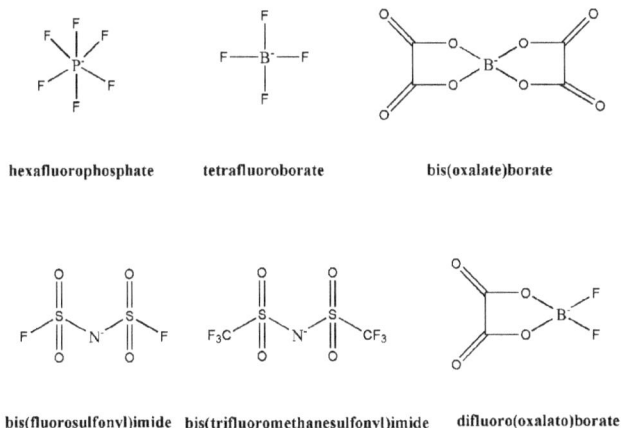

hexafluorophosphate   tetrafluoroborate   bis(oxalate)borate

bis(fluorosulfonyl)imide   bis(trifluoromethanesulfonyl)imide   difluoro(oxalato)borate

Figure 5. Schematic illustration of lithium salts

Because of the small ionic radius of lithium ion, most simple salts of lithium fail to meet the minimum solubility requirement in low dielectric media. Most of the lithium salts that are qualified for the minimal solubility standard are based on complex anions that are composed of a simple anion core stabilized by a Lewis acid agent. anions based on milder Lewis acids can remain stable with organic solvents under normal conditions and have been preferentially investigated by researchers. These salts include lithium perchlorate ($LiClO_4$) and various lithium borates, arsenates, phosphates, and antimonates, $LiMF_n$ (where M = B or As, P, and Sb and n = 4 or 6, respectively). $LiClO_4$ are not thermal stable, $LiAsF_6$ is toxic, the ionic conductivity of $LiBF_4$ is not high enough for practical application, the conductivity of $LiCF_3SO_3$ is not enough, moreover, the serious aluminum corrosion also prevent the application of sulfonate group, Lithium Bis(trifluoromethanesulfonyl)imide is a novel kind of imide based salt for lithium batteries, the larger ion size results in higher ionic conductivity and thermal stability. However, the aluminum corrosion still remains a challenge. Lithium Hexafluorophosphate ($LiPF_6$) the most widely applied lithium salt in the commercial lithium ion batteries. Although $LiPF_6$ is not the best in every single aspect, it has the best balanced properties [46]. In nonaqueous solvents based on mixed alkyl carbonates, $LiPF_6$ remains one of the most conducting salts. For example, in EC/DMC (1:1) the conductivity is 10.7 mS cm^{-1}, it can effectively resist oxidation up to 5.1 V. It also effectively passivates an Al substrate at high potentials by forming a protective film on Al. $LiPF_6$ has become the indispensable electrolyte solute for almost all lithium ion devices. However, there are several disadvantages with $LiPF_6$ salt [46]:

1) chemical and thermal instabilities

$LiPF_6$ could easily decompose even at room temperature with following equation:

$LiPF_6(s) \Leftrightarrow LiF(s) + PF_5(g)$

The PF_5 gas is highly reactive and could initiate a series of reactions, such as ring-opening polymerization of carbonates. Higher temperature will promote the decomposition reaction. Thermal gravimetric analysis (TGA) reveals that, in a dry state, $LiPF_6$ loses 50% of its weight at >200 °C. In nonaqueous solutions, the deterioration occurs at even lower temperatures. This disadvantages limits the high temperature performance of lithium batteries [50].

2) The sensitivity toward ambient moisture

The PF_6 group is sensitive to moisture in the equation by forming LiF, HF with PO_yF_x and loses conductivity, which increases the purification and storage cost. Moreover, The HF could cause the dissolution of metal ion from cathode and results in degradation of battery performance [46,51]. These shortcomings limit the safety and electrochemical performance of lithium ion batteries. Driven by the critical need of EV and HEV, new salts are demanding, great effort has been put on the developing of new salt. Two promising kinds of salts are in vision.

3.1 Boron-based lithium salts

LiBOB has been initially proposed as a possible lithium salt alternative for high temperature operations of LIB and received tensed attention, it has the advantage of moderate conductivity, excellent compatibility with graphite anode, higher thermal stability, stable with Al current collector and low cost [52]. For example, LiBOB could be reduced at 1.7 V (vs. Li/Li^+) and form a stable SEI film, which could enable graphite electrode workable in the PC-based electrolyte [53]. However, its low solubility (<0.8M) and ionic conductivity, anodic unstable when charged to more than 4.2V and high interfacial impedance of the negative electrodes limits its widely application [49, 54].

However, by changing one oxalate group by two F ion in the LiBOB, lithium difluorooxalatoborate (LiDFOB) recently attracted great attention. Lithium difluoro(oxalato)borate (LiDFOB) is first reported in 2006 [54]. It overcomes the electrochemical instability of the LiBOB, while retains the advantages of LiBOB. These advantages enable it to be a promising candidate for lithium ion batteries [54-58].

1) The solubility of LiDFOB is higher than LiBOB, the solubility of LiDFOB in EC/DEC (3:7, w/w) is 1.4 mol/kg compared with 0.65 mol/kg of LiBOB. The transference number of LiDFOB is 0.39 compared with 0.38 of $LiPF_6$.

2) LiDFOB is thermal stable up to 270℃.

3) LiDFOB is resistance to hydrolysis, which could prevent the dissolution of metal ions from cathode materials.

4) Al correct collector is stable up to 4.3V in the electrolyte of 1 M LiDFOB EC/DEC (3:7, w/w), 0.5 V higher than $LiPF_6$.

5) LiDFOB could form more stable SEI films on both anode and cathode electrodes than $LiPF_6$. By adding 2 wt.% LiDFOB, the Li-rich composite cathode capacity retention could be improved to more than 92% after 100 cycles, while the cell loses about 42% of its initial capacity after 100 cycles without LiDFOB. LiDFOB could be reduced on the graphite electrode at 1.6 V (vs. Li/Li^+) during the initial formation process and forms a stable SEI film. The film formation ability of LiDFOB is the higher than FEC.

Amine et al compared the graphite/$Li_{1.1}[Ni_{1/3}Co_{1/3}Mn_{1/3}]_{0.9}O_2$ cells containing lithium bis(oxalate)borate (LiBOB) or vinyl ethyl carbonate (VEC) additives, or the lithium difluoro(oxalato)borate (LiDFOB) additive, they found that the LiDFOB showed the best electrochemical performance compared other additives, The cells were cycled between at 55 °C. The cell without any additive lost about 23% of its initial capacity after 100 cycles, only 5% capacity loss was observed when LiDFOB was used as additive in the electrolyte. LiDFOB could be used as a promising additive to improve the capacity retention, for it is unique in that it does not affect the cell's initial impedance and power [59]. Hu and Bloom further proved that LiDFOB could increase the electrochemical performance of $LiCoPO_4$ and lithium-rich Mn-rich cathodes as LiDFOB can provide an improved SEI film to suppress the degradation of the graphite negative electrode and the cathode [60-61].

Another interesting Boron based lithium salt boron cluster dianions with cage-like structures, the perfluorinated version of which, dilithium dodecafluorododecaborate ($Li_2B_{12}F_{12}$).The large anion size reduce the ion pairing energy in the salt, results in higher solubility. This salt is thermal stable up to 400 ℃, which enabled stable operation of LIB at elevated temperatures. Moreover, this salt could be reversibly oxidized and reduced at around 4.5 V to form an internal short cut current, named "redox shuttle", which could prevent the overcharge of the lithium ion batteries and further increase the safety character. However, the low ionic conductivity may restrict its rate performance. Research demonstrated that by replacing $LiPF_6$ with $Li_2B_{12}F_{12-x}H_x$. Lithium-ion cells using $Li_2B_{12}F_{12-x}H_x$-based electrolytes with proper additives to form stable SEI layer on graphitic anode maintained about 70% initial capacity after cycling at 55℃ for 1,200 cycles. In addition, these cells also showed outstanding intrinsic tolerance towards both continuous and pulse overcharge abuse. In situ high-energy X-ray diffraction study indicated that the delithiated $Li_{1-x}[Ni_{1/3}Mn_{1/3}Co_{1/3}]_{0.9}O_2$ with the presence of $Li_2B_{12}F_{9-x}F_x$ based electrolytes had a better thermal stability than those with the presence of $LiPF_6$-based electrolytes. $Li_2B_{12}F_9H_3$ and $Li_2B_{12}F_{12}$ undergo a reversible redox reaction at 4.5 V and 4.6 V versus Li^+ /Li, respectively, both of which are high enough to provide overcharge protection to 4-V class cathode materials for lithium-ion batteries. When the cell is overcharged, the potential of the cathode reaches the redox potential of $B_{12}F_9H_3^{2-}$ and its shuttle mechanism is then activated, during which $B_{12}F_9H_3^{2-}$ is oxidized (losing an electron) at the cathode and diffuses back to the anode to be reduced (that is, gain an electron from the anode). According to this mechanism, the overcharge current can be shuttled through the lithium-ion cell without further increasing the cell voltage [62, 63].

3.2   LiFSI

Recently, Lithium bis(fluorosulfonyl)imide (LiFSI) received great attention and show promising practical prospect due to its outstanding properties since first reported by Zaghib et al in 2004 [64]. By replacing the CF_3 group in LiTFSI with F we get LiFSI, the LiFSI has following advantages [64-66]:

1) It has even higher solubility and ionic conductivity than $LiPF_6$. 0.85M LiFSI in EC/DMC were found to have a conductivity of 12 mS cm^{-1} at 25 °C. The

conductivity of 1.0 M lithium salt in EC/EMC (3:7, v/v) is 9.73 mS cm^{-1} compared with 9.33 mS cm^{-1} for LiPF$_6$, 7.57 mS cm^{-1} for LiTFSI, 6.26 mS cm^{-1} for LiClO$_4$ and 3.72 mS cm^{-1} for LiBF$_4$.

2) FSI$^-$ anion has a good stability towards hydrolysis. It is reported that the hydrolysis of LiFSI is negligible (<20 ppm) even exposed to water after 22 days.

3) The thermal stability is another advantage of LiFSI, it could be thermal stable up to 200 ℃ and is highly humidity resistance, the temperature at which thermal runway is initiated increases in the following order: LiBF$_4$ > LiPF$_6$ > BETI > LiTFSI > LiFSI.

4) Moreover, it can help forming more stable SEI film. Research showed that LiFSI salt could improve the electrochemical performances of Li/Si cells. PES analysis showed that, unlike LiPF$_6$, due to the low F content of LiFSI and the much lower sensitivity of LiFSI toward hydrolysis, long-term cycling with LiFSI does result neither in the fluorination of the electrode surface to form SiO$_x$F$_y$ nor in the dissolution of the surface lithium oxide Li$_2$O. Instead, a continuous reaction process of SiO$_2$ with lithium upon cycling leads to the increase of lithium silicate Li$_4$SiO$_4$ at the surface of the electrode, which could form a more stable SEI film [66].

Moreover, LiFSI could form a more thermal stable SEI film on the anode. Eshetu et al compared the thermal stabilities and identification of SEI on lithiated graphite in LiFSI and LiPF$_6$ electrolytes. Research found that as LiFSI salt could avoid Lewis acid and lower the HF release upon heating and combustion, lithiated graphite in contact with LiFSI-electrolyte showed a broad but weak peak starting around 80 °C followed by a sharp exotherm around 200 °C. While LiPF$_6$ based electrolyte starts around 120 °C with the SEI break down triggered by an acid-base reaction between the salts precipitated in the SEI (Li$_2$CO$_3$ et al) and the thermally-evolved Lewis acid PF$_5$. These results suggest that LiFSI can hamper the thermal runaway and be of great significance to lithium battery safety [67]. Wang et al compared the affection of lithium bis(oxalato)borate and lithium bis(fluorosulfonyl)imide on lithium manganese oxide spinel lithium-ion batteries. They found that the electrolytes containing 1% LiFSI and 1% LiBOB show better electrochemical stability than the Blank electrolyte, especially for the electrolyte containing 1% LiFSI. High temperature (55 ℃) test showed that LiMn$_2$O$_4$/Li cells with 1% LiBOB delivered ~92.3% 1st CE and improved capacity retention (~82.3% after 100 cycles), and 1% LiFSI delivered ~94.8% 1st CE and improved capacity retention (~85.2% after 100 cycles). While in the absence of additive, the cells show unstable CE during cycling. Moreover, LiFSI salt showed better capacity retention at high current rate, lower Mn dissolution and slower self-discharge rate. However, the Al current collector for the LiFSI-containing electrolyte is slightly corroded. XPS measurement indicates that the addition of LiFSI is help to form a stable and favorable SEI film on the $LiMn_2O_4$ [68].

However LiFSI cannot passivate Al foil very well at higher voltage. Aluminum will be corroded by LiFSI at high potential 3.3 V (vs. Li/Li^+) in LiFSI based carbonate electrolyte as the formed $Al(FSI)_3$ can dissolve in the carbonate based electrolyte. Battery suffers from poor power performance resulting from severe Al corrosion. The passivation of Al can be damaged when trace amounts of LiCl are added into the electrolytes. Lithium difluoro(oxalato)borate (LiDFOB) addition could inhibit the LiFSI corrosion of Al foil by forming of a passive layer composed of Al-F, Al_2O_3, and B-O composition. By substitute one F with nonafluorobutane group, the side reaction of the salt on Al foil could be avoided [69, 70]. Han et al further found that by substituting a F in LiFSI with a nonafluorobutane group, Lithium (fluorosulfonyl) (nonafluorobutanesulfonyl) imide (LiFNFSI) exhibits high thermal stability (thermally stable up to 220 ℃), better cycling performances than $LiPF_6$ and high ionic conductivity. Moreover, LiFNFSI does not corrode Al current collector before 4.5 V vs. Li^+/Li, indicating that LiFNFSI could be promising lithium salt for LIBs [71,72].

4.   Additives

The current state-of-the-art carbonate based electrolyte systems for lithium batteries have drawbacks such as poor solid electrolyte interface (SEI), insufficient cycling performance, safety problems, limited potential window, low ionic conductivity and thermal instability et al. However, till now, the carbonate solvents have the best comprehensive performance. It is a great challenge to develop completely new solvents. Introducing electrolyte additives is one of the most economic and effective methods for the improvement of lithium battery performance. According to parts that the additives play, additives can be divided into: SEI film additives, which can help forming stable SEI films on the cathodes or anodes. Safety additives, which can increase the safety character, of lithium batteries, such as flaming, overcharge and overdischarge. High voltages additives, additives which can improve the solvents decomposition, other additives such as ionic conductivity additives, water adsorption additives are also paid attention to.

4.1   SEI film additives

The electrolytes of lithium batteries are thermodynamically unstable under the potential near the lithium deposition voltage. Hence, a passivation layer named solid electrolyte interphase (SEI) is formed during the first lithiation process, owing to the reduction of electrolytes on the surface of the electrode. The concept of SEI is first put forwarded by Peled in 1979 [73], he proposed that a film is formed instantaneously when lithium metal is contacted with electrolyte. The film is nonconductive to electrons but conductive to ions, the low ionic conductivity makes the diffusion of lithium ions through the SEI to be rate determining step. The films are composed of inorganic components such as $ROCO_2Li$, $(ROCO_2Li)_2$, LiF, Li_2O and Li_2CO_3 et al and organic components such as polycarbonates. The thickness of SEI film is about 2-10 nm, which is ion conductive and electron insulating. Researcher proposed that the SEI might have a multilayer structure within which the inorganic species such as Li_2CO_3 and Li_2O are more stable and closer to lithium, while organic components such as alkyl carbonate are located in the outer layers. An ideal SEI film should meet the following requirements: (1) electron insulator (2) high ion conductivity (3) uniform morphology and chemical composition (4) good mechanical strength and flexibility (5) low solubility in electrolytes [1, 74].

PC is an idea solvent for lithium ion batteries owing to its wide potential window (>5V), high dielectric constant (64.9) and wide temperature range (-49-242℃). However, PC could co-intercalate into the graphite electrode at a potential of ~0.80 V (vs. Li/Li$^+$) and lead to the damage of graphite. This shortcoming prohibits the usage of PC the lithium ion with graphite anode [75]. To use the PC solvent, SEI film additives should be applied. Generally, the LUMO level should be lower than solvents, so the additives can be reduced and form a protective layer on the electrode surface. The additives could be divided into reduction-type additives such as molecules with carbon-carbon double bonds, sulfur-based compounds, and other reductive additives include N_2O, nitrate, nitrite, halogenated ethylene carbonate and halogenated lactone et al. The other kinds is reaction-type additives, such as carboxyl phenol, aromatic esters, anhydride, succinimide et al [76-86]. In 2001, Ogumi et al studied the surface film formation on a graphite negative electrode in lithium-ion batteries via several SEI film additives-vinylene carbonate (VC), fluoroethylene carbonate (FEC), and ethylene sulfite (ES). The reduction potential of VC, FEC, and ES are 1.35, 1.15, and 1.05 V versus Li$^+$/Li, and are all higher than the co-intercalation potential of PC. These additives can be reduced and form a protective layer to cover the whole surface of graphite. It was concluded that the layer of the precipitates functions as a protective surface film, which could suppress cointercalation of PC molecules into the graphite layers. In particular, the addition of VC gave the best charge and discharge characteristics with a high reversible capacity (365 mAh g^{-1}) and good capacity retention (96% at the 50th cycle). In the presence of ES, the Coulombic efficiency in the first cycle was low (64%) because of the presence of a large irreversible capacity. On the other hand, the capacity retention in the presence of FEC was poor, though the initial characteristics were comparable to those in the presence of

VC. The poor capacity retention was due to an electrical isolation of the electrode layer upon prolonged charge and discharge cycling [77].

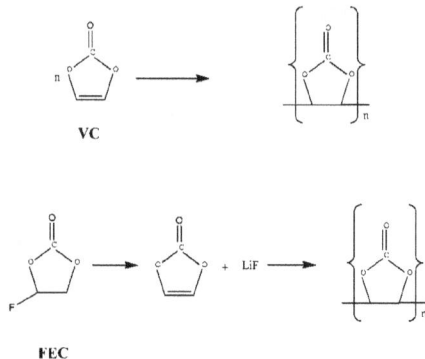

Figure 6. Schematic illustration of SEI film formation of VC and FEC

Abe et al compared a series of olefinic compounds, vinyl acetate (VA), divinyl adipate (ADV) and allyl methyl carbonate (AMC) as the electrolyte additives for PC based LIBs.It is found that 1% of VA, ADV, AMC and ES by weight addition can form SEI films before the reduction of PC.all the olefinic agents have superior effects than ES, as for NG with the highest value of graphitization degree, the cycleability performance ranks in the order of VA, AMC, ADV and ES.The coin cell with the additive ES indicates a sudden drop of discharge capacity versus cycle number, which suggests some poor characteristics of SEI film leading to irreversible loss in capacity. VA appears to be the best agent for the cell cycleabilityas it has following advantage: (1) has lowest impedance, and is thinner than others, (2) possesses the similar SEI film morphology to EC-based electrolyte, and (3) mainly consists of organic compounds and the content of inorganic species like Li and F is very small. They also concluded that the difference of graphitization degree does influence the cycleability, which implies that the quality of SEI film varies with the type of graphitic carbon. With the formation of a proper SEI film by the additive reduction, co-intercalation of the solvent into the graphite is prevented [87].

Silicon is the most promising anode for lithium ion batteries due to its high theoretical capacity (nearly 4200 mAh g^{-1} for $Li_{4.4}Si$). However, the cycling performance of Si is poor due to the large volume expansion of silicon (about 300%). The cracking/crumbling of silicon during cycling and the loss of electrical conduction paths by a large volume change of the active silicon phase during the insertion and extraction of lithium. A continuous SEI-filming process due to continuous crack formation after the first charge-discharge cycle, which exposes new surfaces to the electrolyte will also cause lithium loss. The cycling performance and SEI composition of Si anodes could be improved by the adding of SEI additives.

Xie et al found that the cycle performance and efficiency of Si film anode could be enhanced significantly with the VC addition. The reversible capacity could keep stable even up to 500 cycles. The enhanced electrochemical performance of Si film in VC-containing electrolyte was attributed to the advanced properties of SEI layer formed in initial several cycles. The SEI layer formed in VC-containing electrolyte showed smooth and uniform morphology. Moreover, the presence of VC introduced the VC-reduced products and reduced the LiF content in SEI layer, which resulted in the better conductivity of SEI layer. However, in VC-free electrolyte, the impedance of SEI layer increased constantly upon cycling due to the increasing of SEI layer thickness and high content of LiF, which may cause the increasing of anode polarization and result in the degradation of cycle performance and efficiency of Si film anode [88].

Choi et al probed the fluoroethylene carbonate addition on silicon anode.The reductive decomposition of FEC (LUMO energy: 0.98 eV) is easier than that of EC (1.17 eV). The reduction of FEC progresses before that of EC and thereby an effective SEI layer can be produced at the Si|electrolyte interface.The initial coulombic efficiency of the Si/Li half-cell using the EC/DEC (3/7, v/v) 1.3 M LiPF6 with 3 wt.% FEC as an electrolyte has a higher value of 88.7% and a better capacity retaintion (88.5% vs 67.9% after 80 cycles. The irreversible reaction of blank electrolyte consumes active lithium occurs continuously due to the unstable SEI layer [89].

Lucht et al compared the affection of Fluoroethylene Carbonate and Vinylene Carbonate for Silicon Anodes. They found that 10-15 wt% FEC adding have the small impedance and best capacity retention. The reduction of electrolyte containing FEC forms a stable SEI consisting of poly(EC), LiF, lithium carbonate and lithium alkyl carbonates. Reduction of electrolytes containing VC results in higher impedance and the generation of lithium carbonate, poly(VC) and traces of LiF, and lithium alkyl carbonates [90]. Moreover, Winter et al found that fluoroethylene carbonate (FEC) and vinylene carbonate (VC) could increase the thermal stability of lithiated nano-silicon electrode. The onset temperature of the thermal runaway shifted from 153 ℃ to 200 ℃ and 214 ℃ with 10 wt% FEC and VC addition, respectively [91].

4.2   Safety additives

Lithium ion batteries are now widely used in various portable electronics and also considered as promising candidates for energy storage and Electric Vehicles. However, one key hindrance for further applications of Lithium ion batteries is its safety concern.

Since the state-of-the-art electrolytes of lithium ion batteries use highly flammable and voltage sensitive carbonate based electrolytes, which might causes serious hazards of firing and explosion under abused conditions such as overcharge, heat, crash etc [92].

To solve these problems, much effort has been focused on developing safety additives such as flammable retardants and overcharge inhibitors. Among them, phosphates esters have been proved to be effective as Flame-retardant additives for lithium ion batteries. In the other way, many benzene derivatives were verified as applicable safety additives, which act by forming conducting or isolating polymers inside the batteries to bypass or interrupt the internal current flow [1-6].

Adding flame retarding additives is one of the most economical ways to reduce the flame ability of lithium batteries at the minimized expense of electrochemical performance. A large number of brilliant works have been done on nonflammable additives for lithium batteries since the first report using Trimethyl Phosphate by Wang et al in 2001 [93]. Then fluoro-ether reported by Arai in 2002 [94], fluoro-phosphite reported by Zhang et al in 2003 [95] and phosphonates reported by Yang's and Chen's groups from 2007 [96-99]. The flame retarding mechanism is believed to be the radical-scavenging process, which terminates radical chain reactions responsible for the combustion reaction in the vapor phase when applied for the liquid electrolytes and organic phosphorus flame retarding additives. The first reported TMP is suffered from the poor compatibility with graphite anode as it cannot form SEI films on the graphite electrodes. It is then found that by fluoro substituting the anode compatibility could be greatly improved. The F element could also act as radical terminating agent. When 20 wt.% of tris(2,2,2-trifluoroethyl) phosphate was added, for example, the electrolyte became non-flammable while having no any adverse impacts on both graphite anode and cathode of the Li-ion batteries [94]. Recently, it is found that fluorocyclophosphazene family compounds seem to be a very promising flame retarding additive. The advantages of these compounds include: (1) the increased flame retarding effectiveness due to the high content of phosphorus, nitrogen and fluorine in their ring structure, (2) the excellent stability to the graphite anode and anode. It is reported that addition of 5 wt% PFPN (Ethoxy)pentafluorocyclotriphosphazene can make the electrolyte be totally non-flammable [100]. PFPN is stable even up to 5V (vs. Li/Li^+). The PFPN additive shows a good electrochemical compatibility on the graphitic anode and $LiCoO_2$ cathode. Meanwhile, the incorporated PFPN additive can greatly improve the cyclic performance of $LiCoO_2$ electrode at a high cut-off voltage of 4.5 V. Recently, we found that Pentafluoro(phenoxy)cyclotriphosphazene (FPPN) could not only enable the electrolyte nonflammable with 12% addition but also increase the cycling ability and coulombic efficiency of 5V class $LiNi_{0.5}Mn_{1.5}O_4$ [101].

Since the electrolytes of lithium ion batteries make use of voltage sensitive carbonate based electrolytes, which might causes serious hazards such as firing and explosion under overcharged conditions. As the cells become thermally unstable at overcharged state due to the heat generation, irreversible electrolyte decomposition, highly reactive delithiated cathode and excess lithium metal deposited on the anode, which may ignite or explode finally. To solve these problems, much effort has been focused on developing overcharge additives for 4 V class lithium batteries, such as electropolymerizable additives [102-108], redox shuttles [109-114]. Among them, many benzene derivatives were verified as efficient safety additives, which act by forming polymers inside the batteries to bypass or interrupt the charge current combined with H_2 production to trigger the safety pressure vent. In the commercial electrolyte, overcharge protection additives such as biphenyl (BP) [102] or cyclohexylbenzene (CHB) [104] have been the essential components. Our groups also have proposed a kind of additive with both flame retarding and overcharge protection ability by combining different functional groups [115,116]. Recently, owing to their high energy density, 5 V class cathodes have been considered as promising substitutions for the commercial 4 V class cathodes. Such as $LiNi_{0.5}M_{1.5}O_4$, $LiCoPO_4$ and Lithium-rich cathode ($Li_2MnO_3.LiMO_2$) et al, the working plateau of these cathodes could be as high as 4.8 V [117-129]. However, the working potential of current additives is for 4 V class lithium batteries, it cannot be used in 5 V class lithium batteries due to their low oxidation potential (4.6 V for BP and 4.7 V for CHB). We found that Pentafluoro(phenoxy)cyclotriphosphazene , FPPN, could be used as a safety additive for lithium-ion batteries. It demonstrated that FPPN addition can also be electro-polymerized at 5.05 V (vs. Li/Li^+) and used as an effective overcharge protection additive for 5 V class lithium ion battery. Moreover, FPPN could also be used an efficient flammable retarding additive. The addition of FPPN has positive affection on the electrochemical performance of $LiNi_{0.5}Mn_{1.5}O_4$-Li cells, which suggest it could be a promising additive for lithium ion batteries [101].

Redox shuttle is another kind of overcharge protection additive. It can protect the cell from overcharge reversibly. The shuttle molecules could be oxidized at the cathode and diffuse to the negative electrode and reduced back to the neutral molecule. The redox shuttle acts as an internal shortcut current to bypass the overcharge current and inhibits the cell potential from runaway. The shuttle additive should meet these requirements of: (1) the redox reaction must be highly reversible, (2) its oxidation potential must be slightly higher than the normal operation potential of the positive electrode but lower than the decomposing potential of solvents, (3) it must be electrochemically stable within the cell operating potentials, and (4) its oxidized and reduced forms must be highly soluble and erratic (high diffusion) in the electrolyte. The maximum current that the shuttle additive can carry depends on the concentration of the shuttle molecules in the electrolyte, the diffusion constant of the shuttle molecules, and the number of charges of the shuttle molecules. The redox additives researched earlier is metallocenes, tetracyanoethylene, tetramethylphenylenediamine, dihydrophenazine derivatives and anisole-family compounds, the operation voltage is between 2.8-3.8 V [3]. For example, J.R. Dahn et al reported that 2,5-ditertbutyl-1,4-dimethoxybenzene shows excellent redox shuttle behavior and can provide over 300 cycles of 100% overcharge cycles at 3.9 V versus Li^+/Li, and it is suitable for $LiFePO_4$-based Li-ion batteries, which is too low for commercial 4V class lithium ion batteries [130]. However, this additive can only be used for 3V class lithium ion batteries. Recently, it is found that by incorporating electron withdrawing groups into the benzene ring, the HOMO energy becomes lowered due to the reduced electron density and the shuttle potential could be increased to higher than 4V. For example, 1,4-di-tert-butyl-2,5-bis(2,2,2-trifluoroethoxy)benzene (DBTFB) has a redox potential of 4.25 V, tetraethyl-2,5-di-tert-butyl-1,4-phenylene diphosphate (TEDBPDP) has a potential of 4.75 V vs. Li/Li^+ [131]. However, this modification is not favorable for stabilization of the radical cations that are generated during overcharge, which adversely impacts the electrochemical stability of the shuttle molecules. Not only the radical cations of such high-potential redox shuttle molecules become more energetic and reactive, causing undesired side reactions, but these molecules themselves (due to the presence of these electron withdrawing groups) become more vulnerable to anode reduction, compounding the problem. Only limited cycling performance is obtained. Through the use of strong electron withdrawing organophospine oxide group modification of dimethoxybenzene, K. Amine group found that a reversible redox potential around 4.5 V vs. Li/Li^+ can be obtained and can provide overcharge protection to MCMB/LMO cells for over 25 cycles with 100% overcharge ratio [132]. By introducing electron-withdrawing substituents onto the phenothiazine core, the redox potential can raise its oxidation potential for use as a redox shuttle in high-voltage lithium-ion batteries. N-ethyl-1,2,3,4,6,7,8,9-octafluorophenothiazine could oxidize at 4.3 V vs. Li/Li^+, and can functions for 500 h of 100% overcharge in $LiNi_{0.8}Co_{0.15}Al_{0.05}O_2$/graphite coin cells [133]. Recently, lithium fluorododecaborates ($Li_2B_{12}F_xH_{12-x}$), could serve as a redox shuttle. The doubly charged anion, $(B_{12}F_xH_{12-x})^{2-}$, can be oxidized reversibly at about 4.5 V, which can act as a novel redox shuttle. However, the short cut current is converted to heat to increase the inner cell temperature. High rate overcharge protection may trigger the heat runaway of lithium batteries [134-135].

4.3 Additives for high voltage cathodes

Recently, high voltage cathode materials have been boosting due to its high energy. However, there are two problems that accompanied with the cathode materials.

1) The decomposition of the electrolyte on the cathode materials.

Although most of the electrolyte composition is stable up to 5V (vs. Li/Li$^+$) on the Al foil, however, on the surface of highly oxidative and catalytic activity transition metal oxides, the decomposition potential greatly reduces. The decomposition of carbonates accumulates and will results in poor coulombic efficiency, higher internal resistance and the dissolution of metal ions. EC may be polymerized on the cathodes by cationic mechanisms. Precipitation of surface films on the cathodes, which intensifies at elevated temperature and/or during prolonged storage, increases the electrode's impedance. It is known that LiF films are very resistive to Li-ion migration through them. Surface film formation may also interfere badly with the interparticle electrical contact, and hence, adds internal impedance and problems of electrical contact between the active mass and the current collector. Hence, the apparent capacity fading observed for cathode electrodes during cycling at elevated temperatures and prolonged storage is due to an increase in the electrodes' impedance. One typical example is the lithium-rich cathode. The oxidation degree deteriorates further at elevated temperature. They have poor cycle performance because of the instability of the electrode with the electrolytes, metal dissolution and structural change with the generation of O$_2$ gas.

2) The other problem is the dissolution of metal ions into the electrolytes.

The dissolution of metal ions may cause the degradation of cathode materials. Moreover, the reduction of metal ions on the anode may cause the damages of the SEI film on the anodes. The typical example is LiMn$_2$O$_4$, which is subjected to poor cycling and fast capacity fading. The poor cycling performance can be ascribed to : 1) Disproportionation of Mn^{3+} into Mn^{2+} and Mn^{4+} by HF attack, the Mn^{2+} tends to dissolve in electrolytes and then reduced on the anode to damage the SEI film. 2) The side oxidation reactions between the solvent and the highly oxidative delithiated LiMn$_2$O$_4$. 3) The Jahn-Teller distortion at the end of discharge.

Introducing additives in the electrolyte to form a protection layer on the cathode is proved to be an effective method. A proper additive should satisfying following conditions:

(1) The additive is chemically stable in electrolyte, (2) the additive has lower oxidation potential than those of electrolyte solvents and (3) the additive is decomposed on cathode to form electroconductive member. Concerning about the electrochemical nature, (4) the

additive is required to show no harmful influence on anode in charge–discharge cycles and (5) the resulting electroconductive member has electron conductivity [6].

Aromatic compounds are proved to be effective additives for the cathode materials. Some of these additives can be electropolymerized on the cathode and form a conductive layer to suppress electrolyte decomposition. For example, it is reported that at 0.1 wt.% biphenyl addition leads to the formation of several nm of thin surface layer, which protects further electrolyte decomposition on the cathode active sites. However, the addition should be limited, as oxidative decomposition of the additives progresses as the cycle proceeds, leading to the battery capacity fading because of the grown thick cathode film with high Li^+ resistance. For a 18650 cylindrical cell with $LiCoO_2$ as cathode and graphite as anode, 0.1 wt.% biphenyl addition could greatly improve the cycle ability at 45℃ [137].

Alkene containing compounds are other kinds of effective additives cathode as it can also be electropolymerized at the surface of cathodes. Research found that VC could be cationic polymerized on the delithiated oxides (at potentials >4.2 V vs. Li/Li^+) to form an oligomers surface and suppresses acid-base reactions (e.g. between the $LiMO_2$ and trace HF, PF_5, etc.) that produce resistive surface species such as LiF, Eom et al found that vinylene carbonate (VC) addition could greatly improve the high voltage cycling ability of $LiCoO_2$ 18650 cylindrical cells when charged to 4.35 V. Moreover, VC could improve the high temperature (90℃) cycling performance of $LiCoO_2$. The reason is that VC decomposed on the cathode surface, the oxidation of electrolyte dramatically decreases on the cathode with higher VC concentration via the formation of poly(VC) film on the cathode surface, as suggested by XPS tests [138].

Recently 1,3,5-trihydroxybenzene are also proved as a surface film forming additive on Li-rich layered oxide cathode and demonstrated improved cycling stability of a graphite/$Li_{1.2}Mn_{0.43}Ni_{0.23}Co_{0.23}O_2$ full-cell at 60 °C in the voltage range 2.8-4.3 V. Moreover, trihydroxybenzene could also improve the cycling performance of 5V class $LiNi_{0.5}Mn_{1.5}O_4$ cathode as THB inhibited electrolyte decomposition on the cathode surface [139,140].

Blended additives are also proved to be an effective way, VC together with DFDEC could form an organic/inorganic mixture surface film inhibits metal-dissolution and cathode degradation, and provide high-voltage stability and thermal stability of LMNC cathode material. The cell with LiNMC cathode and graphite anode in a 55 °C under an aggressive charge cut-off voltage to 4.7 V (4.75 V vs. Li/Li^+) showed 77% capacity retention delivering after 50 cycles. However, in conventional electrolyte, expanded structural degradation from surface to bulk, particle cracking, metal dissolution and oxygen loss occur at cathode material, and deposition of metal compounds and structural degradation at graphite anode material, resulting in a rapid capacity fade and very poor capacity retention of 37% at the 50th cycle [141].

Nitrile based additives are lately proved to be another kind of cathode additives for lithium batteries. Succinonitrile is firstly evaluated by Kim et al at 2011 as an additive for improving thermal stability in ethylene carbonate (EC)-based electrolytes for lithium ion batteries. Without any sacrifice of performance such as cyclability and capacity, the introduction of SN into an electrolyte with a graphite anode and Li_xCoO_2 cathode leads to (1) reducing the amount of gas emitted at high temperature, (2) increasing the onset temperature of exothermic reactions and (3) decreasing the amount of exothermal heat. The improvement in the thermal stability is considered to be due to strong complex formation between the surface metal atoms of Li_xCoO_2 and nitrile (CN) groups of SN, By using XPS, evidence of the existence of cobalt-nitrile surface complexes were provided, SN is one of the most potential additives for improving safety, especially for large-scale lithium ion batteries for electric vehicles and energy storage devices [142].

Chen et al found that succinonitrile (SN) could improve the thermal stability and broaden the oxidation electrochemical window of commercial electrolyte for high voltage LIBs. It is found that the $Li_{1.2}Ni_{0.2}Mn_{0.6}O_2$ battery with 1 wt % SN-based electrolyte showed better cyclability and capacity retention when charged to higher cut-off voltage. The improved battery performance is mainly attributed to the formation of uniform cathode electrolyte interface (CEI) formed by interfacial reactions between the LNMO cathode and electrolyte [143].

Recently, Nan et al found that diphenyl disulfide could be used as a new bifunctional electrolyte additive to improve the high-voltage performance of $LiCoO_2$/graphite batteries. 1 wt% DPDS exhibits improved discharge capacity and cycle performance. Research indicates that the DPDS can that the SEI films induced by DPDS can be formed simultaneously on the two electrodes at higher potentials and improve the cyclic performance of battery in the voltage range of 3.0 V-4.4 V [144].

5. Conclusion

The electrolytes are important parts for lithium ion batteries. The need for higher energy brings more demand for novel electrolytes. Further research will focus on safer, wider potential and temperature window electrolytes and more stable SEI films between the electrodes and electrolytes.

Reference

[1]   K. Xu, Chem. Rev.104 (2004) 4303-4417. https://doi.org/10.1021/cr030203g

[2]   R. Fong, U. von Sacken. J.R. Dahn, J. Electrochem. Soc. 137 (1990) 2009. https://doi.org/10.1149/1.2086855

[3]   S.S. Zhang, J. Power Sources 162 (2006) 1379-1394. https://doi.org/10.1016/j.jpowsour.2006.07.074

[4]   A.M. Haregewoin, A. S. Wotango, B.J. Hwang, Energy Environ. Sci. DOI: 10.1039/c6ee00123h. https://doi.org/10.1039/C6EE00123H

[5]   N.S. Choi, J.G. Han, S.Y. Han, I. Park, C.K. Back, RSC Adv., 2015, 5, 2732-2748.

[6]   K. Xu, Chem. Rev. 114 (2014) 11503-11618. https://doi.org/10.1021/cr500003w

[7]   E. Peled, H. Straze, J. Electrochem. Soc. 124 (1977) 1030. https://doi.org/10.1149/1.2133474

[8]   L. Yang, B. Ravdel, B. L. Lucht, Electrochem. Solid-State Lett., 13(2010) A95-A97. https://doi.org/10.1149/1.3428515

[9]   P.G. Balakrishnan, R. Ramesh, T. Prem Kumar, J. Power Sources 155 (2006) 401-414. https://doi.org/10.1016/j.jpowsour.2005.12.002

[10]   L.B. Hu, Z.C. Zhang, K. Amine, Electrochem. Commun. 35 (2013) 76-79. https://doi.org/10.1016/j.elecom.2013.08.009

[11]   M. Inaba, Y. Kawatate, A. Funabiki, S.K. Jeong, T. Abe, Z. Ogumi, Electrochim. Acta 45 (1999) 99-105. https://doi.org/10.1016/S0013-4686(99)00196-6

[12]   J.Arai, H.Katayama, H. Akahoshi, J. Electrochem. Soc. 149 (2002) A217-A226. https://doi.org/10.1149/1.1433749

[13]   J.J. Yuna, L. Zhang, Q.T. Qu, H.m. Liu, X.L. Zhang, M. Shen, H.H. Zheng, Electrochim. Acta 167 (2015) 151-159. https://doi.org/10.1016/j.electacta.2015.03.159

[14]   J.I. Yamaki, I. Yamazaki, M. Egashira, S. Okada, J. Power Sources 102 (2001) 288. https://doi.org/10.1016/S0378-7753(01)00805-9

[15]   Z.C. Zhang, L.B. Hu, H.M. Wu, W. Weng, M. Koh, P.C. Redfern, L.A. Curtiss, K. Amine, Energy Environ. Sci., 6 (2013) 1806-1810. https://doi.org/10.1039/c3ee24414h

[16]   J. Xia, M. Nie, J.C. Burns, A. Xiao, W.M. Lamanna, J.R. Dahn, J. Power Sources 307 (2016) 340-350. https://doi.org/10.1016/j.jpowsour.2015.12.132

[17] J.K. Feng, X.J. Sun, X.P. Ai, Y.L. Cao, H.X. Yang, J. Power Sources 184 (2008) 570-573. https://doi.org/10.1016/j.jpowsour.2008.02.006

[18] J.K. Feng, X.P. Ai, Y.L. Cao, H.X. Yang, J. Power Sources 177 (2008) 194-198. https://doi.org/10.1016/j.jpowsour.2007.10.084

[19] K. Xu, C.A. Angell, J. Electrochem. Soc. 145 (1998) L70. https://doi.org/10.1149/1.1838419

[20] X.G. Sun, C. Austen Angell, Solid State Ionics 175 (2004) 257-260. https://doi.org/10.1016/j.ssi.2003.11.035

[21] X.G. Sun, C. Austen Angell, Electrochem. Commun.7 (2005) 261-266. https://doi.org/10.1016/j.elecom.2005.01.010

[22] L.G. Xue, S.Y. Lee, Z.F. Zhao, C. Austen Angell,J. Power Sources 295 (2015) 190-196. https://doi.org/10.1016/j.jpowsour.2015.06.112

[23] L.G. Xue, K. Ueno, S.Y. Lee, C. Austen Angell, J. Power Sources 262 (2014) 123-128. https://doi.org/10.1016/j.jpowsour.2014.03.099

[24] A. Lewandowski, B. Kurc, I. Stepniak, A. Swiderska-Mocek, Electrochim. Acta 56 (2011)5972-5978. https://doi.org/10.1016/j.electacta.2011.04.105

[25] A. Hofmann, T. Hanemann, J. Power Sources 298 (2015) 322-330. https://doi.org/10.1016/j.jpowsour.2015.08.071

[26] A. Abouimrane, I. Belharouak, K. Amine, Electrochem. Commun. 11 (2009) 1073-1076. https://doi.org/10.1016/j.elecom.2009.03.020

[27] L.P. Mao, B.C. Li, X.L. Cui, Y.Y. Zhao, X.L. Xu, X.M. Shi,S.Y. Li, F.Q. Li, Electrochim. Acta 79 (2012) 197-201. Ionic liquids https://doi.org/10.1016/j.electacta.2012.06.102

[28] P. Hapiot, C. Lagrost, Chem. Rev. 108 (2008) 2238-2264. https://doi.org/10.1021/cr0680686

[29] D.R. MacFarlane, N. Tachikawa, M. Forsyth, J.M. Pringle, P. C. Howlett, G. D. Elliott, J.H. D. Jr., M. Watanabe, P. Simon, C. Austen Angel, Energy Environ. Sci., 7 (2014) 232-250. https://doi.org/10.1039/C3EE42099J

[30] S. Pandian, S.G. Raju, K.S. Hariharan, S. M. Kolake,D. Park, M. Lee, J. Power Sources 286 (2015) 204-209. https://doi.org/10.1016/j.jpowsour.2015.03.130

[31] F. Wu, Q.Z. Zhu, R.J. Chen, N. Chen, Y. Chen,L. Li, Nano Energy 13 (2015) 546-553. https://doi.org/10.1016/j.nanoen.2015.03.042

[32]   J. Towada, T. Karouji, H. Sato, Y. Kadoma, K. Shimada, K. Ui, J. Power Sources 275 (2015) 50-54. https://doi.org/10.1016/j.jpowsour.2014.10.101

[33]   M. Yamagata,N. Nishigaki, S. Nishishita, Y. Matsui, T. Sugimoto, M. Kikuta, T. Higashizaki, M. Kono, M. Ishikaw, Electrochim. Acta 110 (2013) 181-190. https://doi.org/10.1016/j.electacta.2013.03.018

[34]   H.Sakaebe, H. Matsumoto, K. Tatsumi, Electrochim. Acta 53 (2007) 1048-1054. https://doi.org/10.1016/j.electacta.2007.02.054

[35]   G. G. Eshetu, M. Armand, H. Ohno,B. Scrosati, S. Passerini, Energy Environ. Sci., 2016, 9, 49-61. https://doi.org/10.1039/C5EE02284C

[36]   M.V. Fedorov, A.A. Kornyshev, Chem. Rev. 114 (2014) 2978−3036. https://doi.org/10.1021/cr400374x

[37]   T. Nakazawa, A. Ikoma, R. Kido, K. Ueno, K. Dokko, M. Watanabe, J. Power Sources 307 (2016) 746-752. https://doi.org/10.1016/j.jpowsour.2016.01.045

[38]   J.S. Moreno, Y. Deguchi, S. Panero, B. Scrosati, H. Ohno, E. Simonetti, G.B. Appetecchi, Electrochim. Acta 191 (2016) 624-630. https://doi.org/10.1016/j.electacta.2016.01.119

[39]   Y. Matsui, M. Yamagata, S. Murakami, Y. Saito, T. Higashizaki, E. Ishiko, M. Kono, M. Ishikawa, J. Power Sources 279 (2015) 766-773. https://doi.org/10.1016/j.jpowsour.2015.01.070

[40]   H. Sakaebe, H. Matsumoto, Electrochem. Commun. 5 (2003) 594-598. https://doi.org/10.1016/S1388-2481(03)00137-1

[41]   X.G. Sun, S. Dai, Electrochim. Acta 55 (2010) 4618-4626. https://doi.org/10.1016/j.electacta.2010.03.019

[42]   W. Li, J. R. Dahn, D. S. Wainwright, Science 1994, 264, 1115–1118. https://doi.org/10.1126/science.264.5162.1115

[43]   J. Y. Luo, W.-J. Cui, P. He, Y.Y. Xia, Nat. Chem. 2010, 2, 760–765. https://doi.org/10.1038/nchem.763

[44]   J. Y. Luo, Y.Y. Xia, Adv. Funct. Mater. 2007, 17, 3877–3884. https://doi.org/10.1002/adfm.200700638

[45]   R. Ruffo, F.L. Mantia, C. Wessells, R.A. Huggins, Y. Cui, Solid State Ion., 192 (2011) 289–292. https://doi.org/10.1016/j.ssi.2010.05.043

[46]   H. Wang, K. Huang, Y. Zeng, S. Yang, L. Chen, Electrochim. Acta 52 (2007) 3280-3285. https://doi.org/10.1016/j.electacta.2006.10.010

[47] N. Alias, A. Mohamad, J. Power Sources 274 (2015) 237-251.
      https://doi.org/10.1016/j.jpowsour.2014.10.009

[48] L.M. Suo, O. Borodin, T. Gao, M. Olguin, J. Ho, X.L. Fan, C. Luo, C.S. Wang, K.
      Xu, Science 350 (2015) 938-943. https://doi.org/10.1126/science.aab1595

[49] R. Younesi, G. M. Veith, P. Johansson, K. Edstrom, T. Vegge, Energy Environ.
      Sci., 2015, 8, 1905-1922. https://doi.org/10.1039/C5EE01215E

[50] L. Yang, H. Zhang, P. F. Driscoll, B. Lucht and J. B. Kerr, ECS Trans., 33 (2011)
      57-69.

[51] K. Xu, S.S. Zhang, T.R. Jow, W.Xu, C.A. Angell,Electrochem. Solid state lett.5
      (2002) A26-A29. https://doi.org/10.1149/1.1426042

[51] K. Tasaki, K. Kanda, S. Nakamura and M. Ue, J. Electrochem. Soc., 150 (2003)
      A1628-A1636. https://doi.org/10.1149/1.1622406

[52] K. Xu, S.S. Zhang, T.R. Jow, W.Xu, C.A. Angell,Electrochem. Solid state lett.5
      (2002) A26-A29. https://doi.org/10.1149/1.1426042

[53] K. Xu, S.S. Zhang,B.A.Poese, T.R. Jow,Electrochem. Solid state lett.5 (2002)
      A259-A262. https://doi.org/10.1149/1.1510322

[54] S. S. Zhang, Electrochem. Commun., 8 (2006) 1423-1428.
      https://doi.org/10.1016/j.elecom.2006.06.016

[55] Z.H. Chen, J. Liu, K. Amine, Electrochem. Solid state lett.10 (2007) A45-A47.
      https://doi.org/10.1149/1.2409743

[56] T. Schedlbauer, S. Krüger, R. Schmitz, R.W. Schmitz, C. Schreiner, H.J. Gores, S.
      Passerini, M. Winter, Electrochim. Acta 92 (2013) 102-107.
      https://doi.org/10.1016/j.electacta.2013.01.023

[57] S. Zugmann, D. Moosbauer, M. Amereller, C. Schreiner, F. Wudy, R. Schmitz, R.
      Schmitz, P. Isken, C. Dippel, R. Müller, M. Kunze, A. Lex-Balducci, M. Winter,
      H.J. Gores, J. Power Sources 196 (2011) 1417-1424.
      https://doi.org/10.1016/j.jpowsour.2010.08.023

[58] M.Q. Xu, L. Zhou, L.S. Hao, L.D. Xing, W.S. Li, B.L. Lucht, J. Power Sources,
      196 (2011) 6794. https://doi.org/10.1016/j.jpowsour.2010.10.050

[59] J. Liu, Z.H. Chen, S. Busking, I. Belharouak, K. Amine, J. Power Sources 174
      (2007) 852-855. https://doi.org/10.1016/j.jpowsour.2007.06.225

[60] M. Hu, J.P. Wei, L.Y. Xing, Z. Zhou, J. Appl. Electrochem.42 (2012) 291-296.
      https://doi.org/10.1007/s10800-012-0398-0

[61] I. Bloom, L. Trahey, A. Abouimrane, I. Belharouak, X.F. Zhang, Q.L. Wu, W.Q. Lu, D.P. Abraham, M. Bettge, J.W. Elam, X.B. Meng, A.K. Burrell, C.M. Ban, R. Tenent, J. Nanda, N. Dudney, J. Power Sources, 249 (2014) 509-514. https://doi.org/10.1016/j.jpowsour.2013.10.035

[62] Z.H. Chen, Y. Ren, A. N. Jansen, C.K. Lin, W. Weng, K. Amine, Nat. Commun. 4 (2013) 1513. https://doi.org/10.1038/ncomms2518

[63] Z.H. Chen, A.N. Jansen, K. Amine, Energy Environ. Sci., 4 (2011) 4567-4571. https://doi.org/10.1039/c1ee01255j

[64] K. Zaghib, P. Charest, A. Guerfi, J. Shim, M. Perrier, K. Striebel, J. Power Sources, 134 (2004)124-129. https://doi.org/10.1016/j.jpowsour.2004.02.020

[65] H.B. Han, S.S. Zhou, D.J. Zhang, S.W. Fenga, L.F. Li, K. Liu, W.F. Feng, J. Nie, H. Li, X.J. Huang, M. Armand, Z.B. Zhou, J. Power Sources 196 (2011) 3623-3632. https://doi.org/10.1016/j.jpowsour.2010.12.040

[66] B. Philippe, R. Dedryvere, M. Gorgoi, H. Rensmo, D. Gonbeau, K. Edstrom, J. Am. Chem. Soc. 135 (2013) 9829-9842. https://doi.org/10.1021/ja403082s

[67] G.G. Eshetu, S. Grugeon, G. Gachota, D. Mathiron, M. Armand, S. Laruelle, Electrochim. Acta 102 (2013) 133-141. https://doi.org/10.1016/j.electacta.2013.03.171

[68] R.H. Wang, X.H. Li, Z.X. Wang, H.J. Guo, M.R. Su, T. Hou, J. Alloy Compd. 624 (2015) 74-84. https://doi.org/10.1016/j.jallcom.2014.11.098

[69] G.C. Yan, X.H. Li, Z.X. Wang, H.J. Guo, W.J. Peng, Q.Y. Hu, J. Solid state Electrochem. 20 (2016) 507-516. https://doi.org/10.1007/s10008-015-3069-3

[70] S. Zugmann, D. Moosbauer, M. Amereller, C. Schreiner, F. Wudy, R. Schmitz, R. Schmitz, P. Isken, C. Dippel, R. Müller, M.Kunze, A. Lex-Balducci, M. Winter, H.J. Gores, J. Power Sources 196 (2011) 1417–1424. https://doi.org/10.1016/j.jpowsour.2010.08.023

[71] H.B. Han, J. Guo, D.J. Zhang, S.W. Feng, W.F. Feng, J. Nie, Z.B. Zhou, Electrochem. Commun. 13 (2011) 265-268. https://doi.org/10.1016/j.elecom.2010.12.030

[72] L.P. Zheng, H. Zhang, P.F. Cheng, Q. Ma, J.J. Liu, J. Nie, W.F. Feng, Z.B. Zhou, Electrochim. Acta 196 (2016) 169-188. https://doi.org/10.1016/j.electacta.2016.02.152

[73] E. Peled, J. Electrochem. Soc. 126 (1979) 2047. https://doi.org/10.1149/1.2128859

[74] P. Verma, P. Maire 1, P. Novák, Electrochim. Acta 55 (2010) 6332–6341 https://doi.org/10.1016/j.electacta.2010.05.072

[75] M. Arakawa, J. Yamaki, J. Electroanal. Chem. 219 (1987) 273. https://doi.org/10.1016/0022-0728(87)85045-3

[76] S.K. Jeong, M. Inaba, R. Mogi, Y. Iriyama, T. Abe, Z. Ogumi, Langmuir 17 (2001) 8281-8286. https://doi.org/10.1021/la015553h

[77] M.D. Levi, E. Markevich, C. Wang, M. Koltypin, D. Aurbach, J. Electrochem. Soc. 151 (2004) A848. https://doi.org/10.1149/1.1710513

[78] D. Aurbach, K. Gamolsky, B. Markovsky, Y. Gofer, M. Schmidt, U. Heider, Electrochim. Acta 47 (2002) 1423-1439. https://doi.org/10.1016/S0013-4686(01)00858-1

[79] L.B. Hu, Z.C. Zhang, K. Amine, J. Power Sources 236 (2013) 175-180. https://doi.org/10.1016/j.jpowsour.2013.02.064

[80] Y. Luo, T.L. Lu, Y.X. Zhang, L.Q. Yan, J.J. Xie, S.S. Mao, J. Power Sources 323 (2016) 134-141. https://doi.org/10.1016/j.jpowsour.2016.05.053

[81] H.Q. Pham, E.H. Hwang, Y.G. Kwon, S.W. Song, J. Power Sources 323 (2016) 220-230. https://doi.org/10.1016/j.jpowsour.2016.05.038

[82] K. Abe, Y. Ushigoe, H. Yoshitake, M. Yoshio, J. Power Sources 153 (2006) 328-335. https://doi.org/10.1016/j.jpowsour.2005.05.067

[83] Y.S. Kim, T.H. Kim, H. Lee, H.K. Song, Energy Environ. Sci., 2011, 4, 4038-4045. https://doi.org/10.1039/c1ee01272j

[84] S.J. An, J.L. Li, C. Daniel, D. Mohanty, S. Nagpure, D.L. Wood III, Carbon 105 (2016) 52-76. https://doi.org/10.1016/j.carbon.2016.04.008

[85] G. Nagasubramanian, K. Fenton, Electrochim. Acta 101 (2013) 3-10. https://doi.org/10.1016/j.electacta.2012.09.065

[86] S.J. An, J.L. Li, C.Daniel,D. Mohanty, S. Nagpure,D.L. Wood III,Carbon 105 (2016) 52-76. https://doi.org/10.1016/j.carbon.2016.04.008

[87] K. Abe, H. Yoshitake, T. Kitakura, T. Hattoria, H.Y. Wang, M. Yoshio, Electrochim. Acta 49 (2004)4613-4622. https://doi.org/10.1016/j.electacta.2004.05.016

[88] L.B. Chen, K. Wang, X.H. Xie, J.Y. Xie, J. Power Sources 174 (2007) 538-543. https://doi.org/10.1016/j.jpowsour.2007.06.149

[89] N.S. Choi, K.H. Yew, K.Y. Lee, M. Sung, H. Kim, S.S. Kim, J. Power Sources 161 (2006)1254-1259. https://doi.org/10.1016/j.jpowsour.2006.05.049

[90] C.C. Nguyen, B.L. Lucht, J. Electrochem. Soc. 161 (2014) A1933-A1938. https://doi.org/10.1149/2.0731412jes

[91] I.A. Profatilova, C. Stock, A. Schmitz, S. Passerini, M. Winter, J. Power Sources 222 (2013) 140-149. https://doi.org/10.1016/j.jpowsour.2012.08.066

[92] P.G. Balakrishnan, R. Ramesh, T. Prem Kumar, J. Power Sources 155 (2006) 401-411. https://doi.org/10.1016/j.jpowsour.2005.12.002

[93] X.M. Wang, E. Yasukawa, S. Kasuya, J. Electrochem. Soc. 148 (2001) A1058 - A1065. https://doi.org/10.1149/1.1397773

[94] J. Arai, J. Appl. Electrochem. 32 (2002) 1071-1079. https://doi.org/10.1023/A:1021231514663

[95] S.S. Zhang, K. Xu, T.R. Jow, J. Power Sources 113 (2003) 166-172. https://doi.org/10.1016/S0378-7753(02)00537-2

[96] H.F. Xiang, Q.Y. Jin, C.H. Chen, X.W. Ge, S. Guo, J.H. Sun, J. Power Sources 174 (2007) 335-341. https://doi.org/10.1016/j.jpowsour.2007.09.025

[97] J.K. Feng, X.J. Sun, X.P. Ai, Y.L. Cao, H.X. Yang, J. Power Sources 184 (2008) 570-573. https://doi.org/10.1016/j.jpowsour.2008.02.006

[98] Z.Q. Zeng, X.Y. Jiang, B.B. Wu, L.F. Xiao, X.P. Ai, H.X. Yang, Y.L. Cao, Electrochem. Acta 129 (2014) 300-304. https://doi.org/10.1016/j.electacta.2014.02.062

[99] H.F. Xiang, Q.Y. Jin, C.H. Chen, X.W. Ge, S. Guo, J.H. Sun, J. Power Sources 174 (2007) 335-341. https://doi.org/10.1016/j.jpowsour.2007.09.025

[100] L. Xia, Y.G. Xia, Z.P. Liu, J. Power sources 278 (2014) 190-196. https://doi.org/10.1016/j.jpowsour.2014.11.140

[101] J.K. Feng, X.P. Gao,L.J. Ci, S.L. Xiong, RSC Adv., 6 (2016) 7224-7228.

[102] L.F. Xiao, X.P. Ai, Y.L. Cao, H.X. Yang, Electrochim. Acta 49 (2004) 4189-4196. https://doi.org/10.1016/j.electacta.2004.04.013

[103] S.L. Li, X.P. Ai, J.K. Feng, Y.L. Cao, H.X. Yang, J. Power Sources 184 (2008) 553-556. https://doi.org/10.1016/j.jpowsour.2008.02.041

[104] M.Q. Xu, L.D. Xing, W.S.Li, X.X. Zuo, D. Shu, G.L. Li, J. Power Sources 184 (2008) 427-431. https://doi.org/10.1016/j.jpowsour.2008.03.036

[105] N. Iwayasu, H. Honboua, T. Horiba, J. Power Sources 196 (2011) 3881-3886. https://doi.org/10.1016/j.jpowsour.2010.12.082

[106] Y.G. Lee, J. Cho, Electrochim. Acta 52 (2007) 7404-7408. https://doi.org/10.1016/j.electacta.2007.06.032

[107] B. Wang, Q. Xia, P. Zhang, G.C. Li, Y.P. Wu, H.J. Luo,S.Y. Zhao, T. Van Ren, Electrochem. Comm. 10 (2008) 727-730. https://doi.org/10.1016/j.elecom.2008.02.011

[108] X.M. Feng, X.P. Ai, H.X. Yang, J. Appl. Electrochem. 34 (2004) 1199-1203. https://doi.org/10.1007/s10800-004-0771-8

[109] L.M. Moshuchak, M. Bulinski, W.M. Lamanna, R.L. Wang, J.R. Dahn. Electrochem. Commun. 9 (2007) 1497-1501. https://doi.org/10.1016/j.elecom.2007.01.059

[110] J.H. Huang, I. A. Shkrob, P.Q. Wang, L. Cheng, B.F. Pan, M.N. He, C. Liao, Z.C. Zhang, L. A. Curtiss, L. Zhang, J. Mater. Chem. A 3 (2015) 7332-7337 https://doi.org/10.1039/C5TA00899A

[111] J.K. Feng, X.P. Ai, Y.L. Cao, H.X. Yang, Electrochem. Commun. 9 (2007) 25-30. https://doi.org/10.1016/j.elecom.2006.08.033

[112] C.M. Ionica-Bousquet, D. Mu-oz-Rojas, W.J. Casteel, R.M. Pearlstein, G. GirishKumar, G.P. Pez, M.R. Palacín, J. Power Sources 195 (2010) 1479-1485. https://doi.org/10.1016/j.jpowsour.2009.09.023

[113] J.W. Wen, D.W. Zhang, C.H. Chen, C.X. Ding, Y. Yu, J. Maier, J. Power Sources 264 (2014) 155-160. https://doi.org/10.1016/j.jpowsour.2014.04.077

[114] J. Lamb, C.J. Orendorff, K. Amine, G. Krumdick, Z.C. Zhang, L. Zhang, A.S. Gozdz, J. Power Sources 247 (2014) 1011-1017. https://doi.org/10.1016/j.jpowsour.2013.08.044

[115] J.K. Feng, Y.L. Cao, X.P. Ai, H.X. Yang, Electrochim. Acta 53 (2008) 8265-8268. https://doi.org/10.1016/j.electacta.2008.05.024

[116] J.K. Feng, L. Lu, J. Power Sources 243 (2013) 29-32. https://doi.org/10.1016/j.jpowsour.2013.05.170

[117] J.H. Kim, N.P.W. Pieczonka, L. Yang, Chemphyschem 15 (2014) 1940-1954. https://doi.org/10.1002/cphc.201400052

[118] A. Kraytsberg, Y. Ein-Eli, Adv. Energy Mater. 2 (2012) 922-939. https://doi.org/10.1002/aenm.201200068

[119] L. Li, K.S. Lee, L. Lu, Funct. Mater. Lett. 7 (2014) 1430002.
https://doi.org/10.1142/S1793604714300023

[120] J. Kalhoff, G.G. Eshetu, D. Bresser, S. Passerini, Chemsuschem 8 (2015) 2154-2175. https://doi.org/10.1002/cssc.201500284

[121] N.S. Choi, J.G. Han, S.Y. Ha, I. Park, C.K. Back, RSC Adv. 5 (2015) 2732-2748.

[122] S. Tan, Y.J. Ji, Z.R. Zhang, Y. Yang, Chemphyschem 15 (2014) 1956-1969.
https://doi.org/10.1002/cphc.201402175

[123] M. Hu, X.L. Pang, Z. Zhou, J. Power Sources 237 (2013) 229-242
https://doi.org/10.1016/j.jpowsour.2013.03.024

[124] N.S. Choi, Z.H. Chen, S.A. Freunberger, X.L. Ji, Y.K. Sun, K. Amine, G. Yushin, L.F. Nazar, J. Cho, P.G. Bruce, Angew. Chem. Int. Ed. 51 (2012) 9994-10024.
https://doi.org/10.1002/anie.201201429

[125] J. Lu, Y.L. Chang, B.H. Song, H. Xia, J.R. Yang, K.S. Lee, L. Lu, J. Power Sources 271 (2014) 604-613. https://doi.org/10.1016/j.jpowsour.2014.08.037

[126] J. Zou, L. Lu, Mater. Tech. 30 (2015) A1-A1.
https://doi.org/10.1179/A15Z.0000000009

[127] X. Tang, S.S. Jan, Y.Y. Qian, H. Xia, J.F. Ni, S.V. Savilov, S.M. Aldoshin, Sci. Rep. 5 (2015) 11958. https://doi.org/10.1038/srep11958

[128] J.F. Ni, W. Liu, J.Z. Liu, L.J. Gao, J.T. Chen, Electrochem. Comm. 35 (2013) 1-4.
https://doi.org/10.1016/j.elecom.2013.07.030

[129] H.M.Cho, M.V. Chen, A.C. MacRae, Y.S. Meng, ACS. Appl. Mater. Inter. 7 (2015) 16231-16239. https://doi.org/10.1021/acsami.5b01392

[130] J. Chen, C. Buhrmester, J.R. Dahn, Electrochem.Solid-State Lett., 8 (2005) A59-A62. https://doi.org/10.1149/1.1836119

[131] Z.Q. Zeng, X.Y. Jiang, B.B. Wu, L.F. Xiao, X.P. Ai, H.X. Yang, Y.L. Cao, Electrochem. Acta 129 (2014) 300-304.
https://doi.org/10.1016/j.electacta.2014.02.062

[132] J.H. Huang, N. Azimi, L. Cheng, I. A. Shkrob, Z. Xue, J.J. Zhang, N.L. Dietz Rago, L.A. Curtiss, K. Amine, Z.C. Zhang, L. Zhang, J. Mater. Chem. A, 2015, 3, 10710-10714. https://doi.org/10.1039/C5TA01326G

[133] A.P. Kaur, M.D. Casselman, C.F. Elliott, S.R. Parkin, C. Riskoab, S.A. Odom, J. Mater. Chem. A, 4 (2016) 5410-5414. https://doi.org/10.1039/C5TA10375D

[134] Z.H. Chen, A.N. Jansen, K. Amine, Energy Environ. Sci., 4 (2011) 4567-4571. https://doi.org/10.1039/c1ee01255j

[135] Z. Chen, J. Liu, A.N. Jansen, G. GirishKumar, B. Casteel, K. Amine, Electrochem.Solid-State Lett., 13 (2010) A39-A42. https://doi.org/10.1149/1.3299251

[136] A. M. Haregewoin, A.S. Wotango, B.J. Hwang, Energy Environ. Sci.9 (2016) 1955-1988. https://doi.org/10.1039/C6EE00123H

[137] K. Abe, Y. Ushigoe, H. Yoshitake, M. Yoshio, J. Power Sources 153 (2006) 328-335. https://doi.org/10.1016/j.jpowsour.2005.05.067

[138] J.Y. Eom, I.H. Jung, J.H. Lee, J. Power Sources 196 (2011) 9810-9814. https://doi.org/10.1016/j.jpowsour.2011.06.095

[139] A. Perea, K. Zaghib, D. Belanger, J. Mater. Chem. A 3 (2015) 2776-2783. https://doi.org/10.1039/C4TA05767H

[140] Y.S. Kang, T. Yoon, J. Mun, M.S. Park, I.Y. Song, A. Benayad, S.M. Oh, J. Mater. Chem. A 2 (2014) 14628-14633. https://doi.org/10.1039/C4TA01891E

[141] H.Q. Pham, E.H. Hwang, Y.G. Kwon, S.W. Son, J. Power Sources 323 (2016) 220-230. https://doi.org/10.1016/j.jpowsour.2016.05.038

[142] Y.S. Kim, T.H. Kim, H. Lee, H.K. Song, Energy Environ. Sci., 4 (2011) 4038-4045. https://doi.org/10.1039/c1ee01272j

[143] R.J. Chen, F. Liu, Y. Chen, Y.S. Ye, Y.X. Huang, F. Wu, L. Li, J. Power Sources 306 (2016) 70-77. https://doi.org/10.1016/j.jpowsour.2015.10.105

[144] M.K. Zhao, X.X. Zuo, X.D. Ma, X. Xiao, L. Yu, J.M. Nan, J. Power Sources 323 (2016) 29-36. https://doi.org/10.1016/j.jpowsour.2016.05.052

CHAPTER 7

Garnet-type Li Ion Conductive Ceramics and its Application for All-solid-state Li Batteries

M. Kotobuki

Department of mechanical engineering, National University of Singapore

mpemako@nus.edu.sg

Abstract

Li ion conductive ceramics are expected to be a solid electrolyte for all-solid-state Li batteries which could resolve safety issues in present Li batteries. The Li ion conductive ceramics can be in general categorized into oxide and sulfide groups. The sulfide-based ceramics show high Li ion conductivity but are not stable in air and produces toxic H_2S gas. In contrary, oxide-based ceramics are stable in air. Therefore, the oxide-based ceramics are more favorable to manufacture. In this chapter, garnet-type Li ion conductive ceramics which are one of the most widely studied oxide-based Li ion conductive ceramics are reviewed in detail.

Keywords

Li Ion Conductive Ceramics, Solid Electrolyte, All-Solid-State Battery, Garnet, Lithium-Ion Battery

Contents

1. Introduction

Ion conductive ceramics have attracted a lot of attention for a wide variety of electrochemical applications, ranging from power generation (solid oxide fuel cell, SOFC) and energy storage (battery and capacitor) to atomic switches [1]. The ion conductive ceramics are sometimes called "solid electrolyte" when they are used for electrochemical devices.

Among them, Li ion conductive ceramics are expected to be a solid electrolyte for all-solid-state Li batteries which can solve safety issues in present Li batteries. Also, the Li ion conductive ceramics can be used as a Li ion conductive membrane for Li-air batteries which can meet the increasing demands for energy storage devices with high energy density [2]. Li ion conductive ceramics can be categorized into crystalline and amorphous (glass) phases and glass-ceramics phase which is a mixture of crystalline and amorphous phases (Fig. 1). The crystalline ceramics is composed of rigid skeleton through which Li ion migrates with a low activation energy. The Li ion conductivity of the crystalline

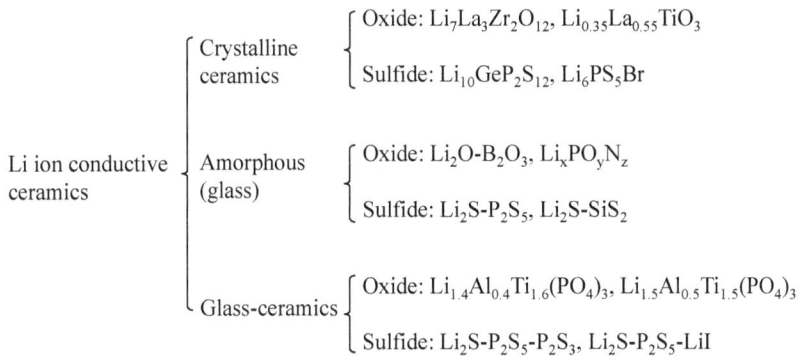

Li ion conductive ceramics
- Crystalline ceramics
  - Oxide: $Li_7La_3Zr_2O_{12}$, $Li_{0.35}La_{0.55}TiO_3$
  - Sulfide: $Li_{10}GeP_2S_{12}$, Li_6PS_5Br
- Amorphous (glass)
  - Oxide: $Li_2O\text{-}B_2O_3$, $Li_xPO_yN_z$
  - Sulfide: $Li_2S\text{-}P_2S_5$, $Li_2S\text{-}SiS_2$
- Glass-ceramics
  - Oxide: $Li_{1.4}Al_{0.4}Ti_{1.6}(PO_4)_3$, $Li_{1.5}Al_{0.5}Ti_{1.5}(PO_4)_3$
  - Sulfide: $Li_2S\text{-}P_2S_5\text{-}P_2S_3$, $Li_2S\text{-}P_2S_5\text{-}LiI$

Figure 1. Categories of Li ion conductive ceramics.

ceramics is generally 1 to 2 orders higher than that of amorphous ones. Each category can be further classified into oxide and sulfide groups. Normally, sulfide-based Li ion conductive ceramics exhibits higher Li ion conductivity than those of oxide-based ones because the high polarization of sulfide ions weakens the interaction between the anions and the lithium ions. Additionally, a good contact among crystal grains in sulfide-based ceramics can be obtained easily by cold-pressing because of their elastic nature [3]. Due to these promising natures of the sulfide-based ceramics, characteristics of their ionic conductions, crystallography of structures and construction of the all-solid-state batteries based on sulfide electrolyte have been studied, although the sulfide-based ceramics react with ambient moisture and generate toxic H_2S gas [4]. Due to instability of the sulfide ceramics in air, especially in moisture, all production processes must be done in dry condition. This is a large shortcoming of the sulfide ceramics for practical application. Contrary, oxide-based ceramics are more stable in air and high temperature. In addition, precursors used for the oxide-based ceramics can be obtained and stored easily. Therefore, the oxide-based ceramics are more favorable to manufacture. The Li ion conductivity of the oxide-based ceramics is usually one order of magnitude lower than that of sulfide-based ones. Despite the drawback with respect to the conductivity, the oxide-based Li ion conductive ceramics have received much attention for practical use because of their stability in air, ease of handling and low production cost. In this chapter, garnet-type Li ion conductive ceramic which is one of the most widely researched oxide-based Li ion conductive ceramics is reviewed.

2. $Li_5La_3M_2O_{12}$ (M=Nb, Ta) Ceramics

Li ion garnet compounds that are expressed in the general formula $Li_5La_3M_2O_{12}$ (M = Ta, Nb) was firstly shown by Thangadurai et. al. in 2003 [5]. The Li garnet, $Li_3Ln_3Te_2O_{12}$ shows low Li ion conductivity ($\sim10^{-5}$ S cm^{-1} at 600 °C) [6]. In contrary, Li-stuffed garnet that is defined by more than three Li per formula unit such as $Li_5La_3Ta_2O_{12}$ and $Li_5La_3Nb_2O_{12}$ shows about three order of magnitude higher Li ion conductivity ($\sim10^{-5}$ S cm^{-1} at room temperature) [5, 7]. This finding provided an impact for a development of the garnet-type Li ion conductive ceramics. Since then, the garnet-type Li ion conductive ceramics are intensively studied by many research groups.

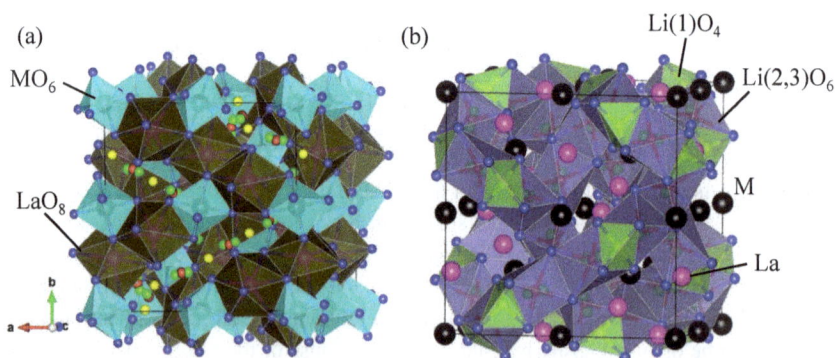

Fig. 2 Structure of $Li_5La_3M_2O_{12}$ (M=Nb, Ta): (a) MO_6 and LaO_8 polyhedra with yellow spheres representing 24d (Li1) site, green ones 96h (Li2) site and blue ones 48g (Li3) site, and (b) LiO_4 and LiO_6 polyhedra.

A crystal structure of $Li_5La_3M_2O_{12}$ is shown in Fig. 2. The $Li_5La_3M_2O_{12}$ crystal belongs to the space group of *Ia-3d*. The framework is composed of octahedral MO_6 and dodecahedral LaO_8. Li ions occupy the tetrahedral 24d site and distorted octahedral 96h/48g sites. In $Li_3Ln_3Te_2O_{12}$, Li ions only exist in the 24d site and are tightly bound in the 24d sites. Therefore, the Li ion conductivity of $Li_3Ln_3Te_2O_{12}$ is low. In $Li_5La_3M_2O_{12}$, Li ions occupy in the tetrahedral 24d site ($Li(1)O_4$) and distorted octahedral 96h/48g sites ($Li(2,3)O_6$). Both 24d and 96h/48g sites are partially occupied [8], leading to Li vacancies in the tetrahedral sites.

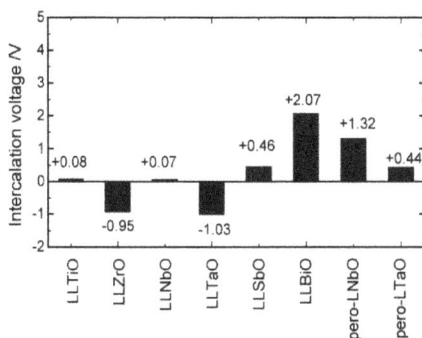

Figure 3. Li intercalation voltage of various Li ion conductive ceramics calculated by First-principle density functional theory.

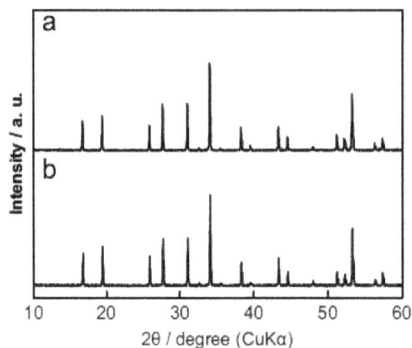

Figure 4. XRD patterns of LLTaO (a) before and (b) after contacting melting Li metal for 72 h.

$Li_5La_3Ta_2O_{12}$ (LLTaO) is considered to be one of the promising solid electrolytes for all-solid-state battery. LLTaO has been confirmed to be stable in contact with Li metal theoretically and experimentally [9]. Fig. 3 reveals intercalation voltage of various garnet ceramics calculated by first-principle density functional theory. The intercalation voltage of LLTaO has a negative value, implying that LLTaO is stable in contact with Li metal. Fig. 4 depicts XRD patterns of LLTaO before and after contacting Li metal. Both patterns were identical [10]. This means Li metal can be used as anode material for the all-solid-state battery with LLTaO solid electrolyte. Theoretical capacity of Li metal anode (3862 mAh g^{-1}) is about 10 times higher than that of a graphite anode (372 mAh g^{-1}) used in the commercial Li batteries [11]. Therefore, anode volume can be reduced to about one-tenth if the graphite anode can be replaced by a Li metal anode, leading to high energy density of the all-solid-state battery. Furthermore, it was reported that LLTaO did not react with $LiCoO_2$ [10]. Fig. 5 shows XRD patterns of LLTaO impregnated precursor sol for $LiCoO_2$ followed by calcination at 700 °C for 2h. All diffraction peaks can be attributed to $LiCoO_2$ and LLTaO. No impurity phase was observed.

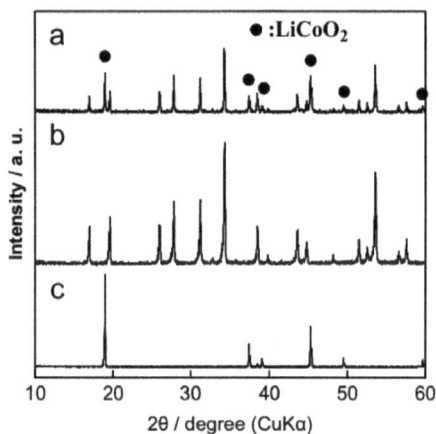

Figure 5. XRD patterns of (a) LLTaO after impregnation of $LiCoO_2$, (b) LLTaO and (c) $LiCoO_2$.

Cyclic voltammograms of Li/LLTaO/LiCoO$_2$ cell is exhibited in Fig. 6. In the anode scan, two distinct peaks by oxidation were observed at 3.75 and 3.95 V vs. Li/Li$^+$ which corresponded to oxidation of LiCoO$_2$ [12]. In the cathode scan, reduction peaks were observed as well. Surprisingly, the redox peaks were still confirmed even after one year of storage of the cell in a Ar-filled globe-box, indicating that the cell was still working (Fig. 6b). It is concluded that the interfaces of LiCoO$_2$/LLTaO and Li/LLTa are extremely stable. These features prove LLTaO is a promising solid electrolyte for the all-solid-state battery with Li metal anode.

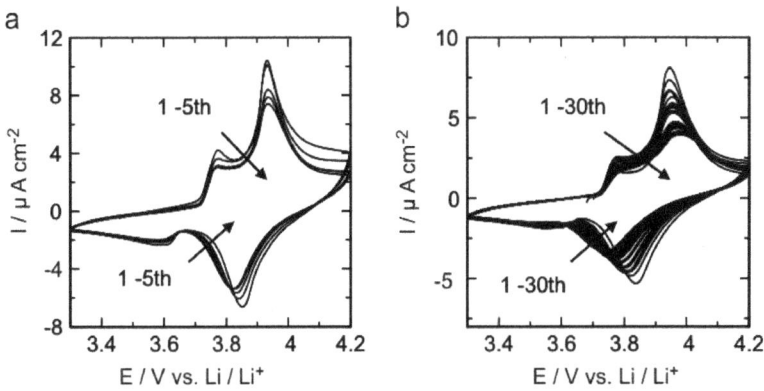

Figure 6. Cyclic voltammogram of Li/LLTaO/LiCoO$_2$ cell: (a) just after fabrication, and (b) after preservation in Ar-filled globe-box for a year.

To improve the Li ion conductivity, hetreroatom substitution has been performed. M (=Nb and Ta) in Li$_5$La$_3$M$_2$O$_{12}$ can be replaced with Bi [13, 14], Sb[15, 16] and In [17]. La also can be substituted by Pr, Nd [18], Ba [19] Ca [20] and K [17]. Structural parameter and Li ion conductivity of these garnets are summarized in literature [21].

3. $Li_7La_3Zr_2O_{12}$ Ceramics

In 2007, the discovery of $Li_7La_3Zr_2O_{12}$ (LLZ) opened a new field of garnet-type Li ion conductive ceramics [22]. The Li ion conductivity of LLZ was reported to 2.44×10^{-4} S cm^{-1}. LLZ also showed good chemical and electrochemical stabilities against Li metal and wide electrochemical window [23]. Fig. 7 reveals a cyclic voltammogram of the Li/LLZ/Li cell. The dissolution and deposition reactions of Li metal were observed reversibly, indicating that Li ion could transport through the LLZ without any degradation of LLZ.

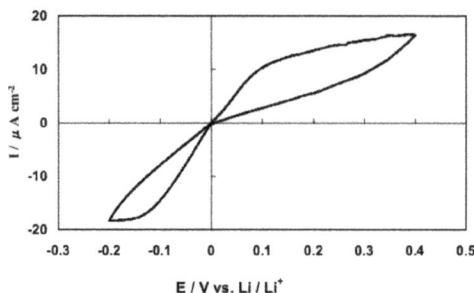

Figure 7. Cyclic voltammogram of Li/LLZ/Li cell.

Fig. 8 depicts a structure of LLZ. LLZ has a cubic symmetry with space group *Ia-3d* and the cubic lattice parameter was determined to be 12.9827(4) Å using single crystal data [24]. The framework of LLZ is the same as $Li_5La_3M_2O_{12}$, i.e. the framework with dodecahedral LaO_8 and octahedral ZrO_6. The difference is the amount of Li ion. In LLZ, seven Li ions are involved per formula unit. When the Li ions occupy both neighboring tetrahedral 24d and octahedral 48g/96h sites, the Li ions in the octahedral sites move from the 48g position to 96h position by the Coulomb repulsion among the neighboring Li ions [25] (Fig. 9). This displacement is a reason for the high ionic conductivity of LLZ. Fig. 10 reveals a relationship between Li ion occupancy in different coordinate sites and ionic conductivity as a function of Li content in the LLZ. It can be clearly seen that the ionic conductivity increases in Li content and Li occupancy in 48g/96h sites [26]. Therefore, LLZ containing seven Li ions in the formula unit exhibits about two order higher ionic conductivity than LLTaO with only five Li ions in the formula unit.

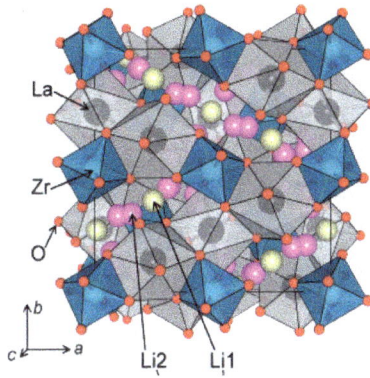

Figure 8. Structure of LLZ.

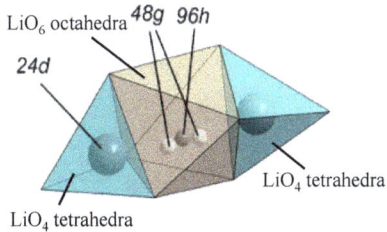

Figure 9. Li sites in LLZ.

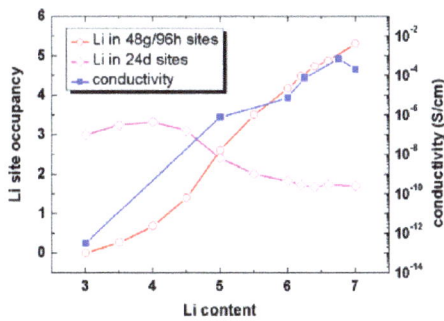

Figure 10. Relationship between Li occupancy in various Li sites and conductivity.

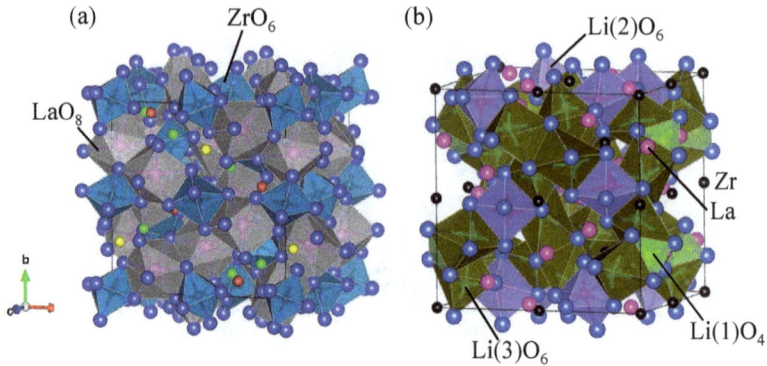

Figure 11. Structure of tetragonal LLZ. Two structures are same. Difference is expression of polyhedral.(a) ZrO_6 and LaO_8 polyhedra, Yellow: 8a (Li1) site, Green: 96h16f (Li2) site, Blue: 32g (Li3) site, (b) LiO_4 and LiO_6 polyhedra.

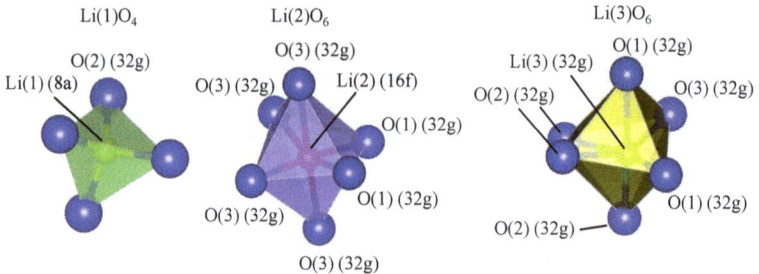

Figure 12. Structures of Li(1)O_4, Li(2)O_6 and Li(3)O_6 in tetragonal LLZ.

Although LLZ is an attractive material for a solid electrolyte, the presence of tetragonal in LLZ shows about two order lower in Li ion conductivity ($\sim 10^{-6}$ S cm^{-1}) than that of the cubic phase. The tetragonal phase is thermodynamically more stable than that of the cubic phase at room temperature [27]. The crystal structure of tetragonal LLZ is depicted in Fig. 11 [28]. The tetragonal LLZ is expressed in space group of $I4_1/acd$. The lattice parameters are a=13.134(4) Å and c=12.663(8) Å. The garnet framework in the tetragonal LLZ is composed of dodecahedral La(1)O_8 and La(2)O_8 and octahedral ZrO_6. Li ions occupy three sites. Li(1) ions occupy tetrahedral 8a sites. Li(2) and Li(3) ions sit in octahedral 16f and 32g sites, respectively (Fig.12)[28]. The tetrahedral Li(1)O_4 is

172

composed of only O(2) (32g) oxygen. Contrary, the octahedral $Li(2)O_6$ consists of O(1) (32g) and O(2) (32g) oxygens. The distorted octahedral $Li(3)O_6$ is located between $Li(1)O_4$ tetrahedra and $Li(2)O_6$ octahedra and is comprised of O(1), (2) and O(3) (32g) oxygens. In the tetragonal LLZ, the Li(1), Li(2) and Li(3) sites are fully occupied. Contrary, there are only two Li sites (tetrahedral 24d site and distorted octahedral 96h sites) that exist in the cubic phase. These sites are not fully occupied and some Li vacancies exist. Fig. 13 depicts Li ion arrangement of cubic and tetragonal LLZ [24]. In the cubic phase, the Li ions build a loop and the loop is shared with other loops at the Li(1) site. There are two Li(2) sites between the Li(1) sites and either Li(2) sites would be occupied because these two Li(2) sites are too close to each other. The loops also exist in the tetragonal phase, but the Li ion arrangement is different. The Li(1) site is shared with other loops as well, however, one Li(2) and two Li(3) exist between the Li(1) sites and construct the loop arrangement of Li(1)-Li(3)-Li(2)-Li(3)-Li(1). The distance between Li(1) and Li(2) sites in the cubic phase is 1.602(18) Å. On the other hand, in the tetragonal phase, the Li-Li distance is longer, > 2.5 Å. The short Li-Li distance in the cubic phase derives the Coulomb repulsion and then increases the mobility of Li ions, resulting in high Li ion conductivity of the cubic phase. Recently, the third phase of LLZ was found. This phase is called low temperature(LT)-cubic phase [29]. The LT-cubic phase is formed when the tetragonal LLZ is heated to about 100 °C. Li_2CO_3 formation due to absorption of CO_2 in air by LLZ and simultaneously generation of Li vacancies are thought to be a reason for the formation of LT-cubic phase [30]. However, the reason for LT-cubic formation is still being debated.

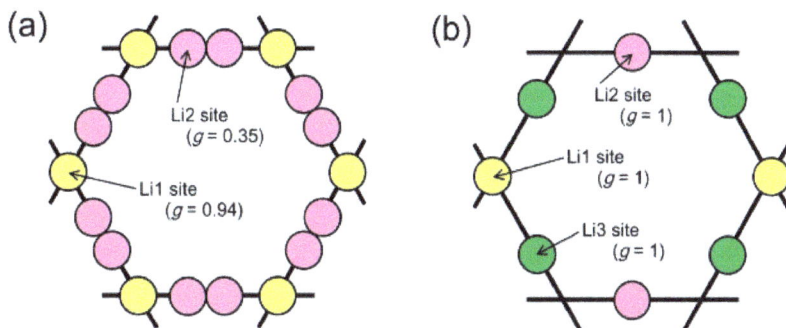

Figure 13. Li ion arrangements in (a) cubic and (b) tetragonal phases.

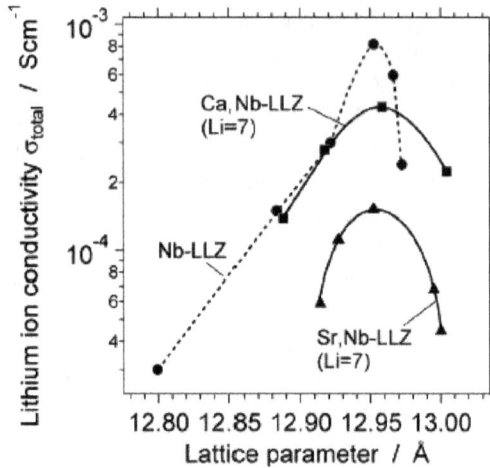

Figure 14. Li ion conductivity and lattice constant of Nb-LLZ, Sr, Nb-LLZ and Ca,Nb-LLZ.

The tetragonal LLZ requires full Li ion occupancy. To obtain a high Li ion conductive cubic phase at ambient temperature, a creation of Li vacancies is needed. A critical Li vacancy concentration for the stabilization of cubic phase was predicted 0.4-0.5 per formula unit by a combination of density-functional theory and molecular dynamics simulations [31]. To create the Li vacancies, substitution with foreign multivalent atoms has been widely performed. As substitution elements, Al [32-35], Ta [27, 36, 37], Ga [38, 39], Ge [40], Nb [41], Te [42], Y [43], Cr [44] and so on has been attempted and cubic LLZ was obtained successfully. The substitution sites are different in different substituents. Al, Ga and Ge are considered to enter into the Li sites and Ta, Y, Cr, Nb and Te exist in the Zr site. This indicates that the substitution sites to create the Li vacancies does not relate to the stabilization of cubic phase. However, the Li site substitution would block Li ion conduction by the substituent. The Li ion conductivity of Al-free cubic phase at 25 °C was estimated to 1.71×10^{-3} S cm^{-1} by extrapolating the Li ion conductivity at 650 ~850 °C [45], which is about one order higher than the Al-doped cubic phase. By structural analysis from neutron diffraction, the maximum Li content in the cubic garnet would be 7.5 per formula unit and maximum conductivity would appear at Li content of 6.4±0.1 per formula [46]. Lattice constant is also thought to influence the Li ion conductivity of cubic LLZ. Optimal lattice constant for the Li ion conduction in

Li_{7+x-y} $(La_{3-x}A_x)(Zr_{2-y}Nb_y)O_{12}$ (A = Alkaline metal) was suggested to 12.94 ~ 12.96 Å (Fig. 14) [47].

However, reproducibility, especially with respect to Li ion conductivity of LLZ is a critical issue for the development of LLZ. Ga, Ta or Te substituted LLZ showed the Li ion conductivity of ~ 10^{-3} S cm^{-1} [42, 48-50], but lower conductivity value was also reported in similar compositions [38, 51, 52]. This discrepancy is still on debate, but deviation of sample composition is suggested as the reason. The Li ion conductivity of LLZ is strongly influenced by Li content in the sample. LLZ is normally exposed in high temperature sintering (> 1200 °C) to obtain a sintered LLZ body. At this high temperature, Li evaporation from the sample occurs. A precise control of the evaporation is very difficult, leading to the deviation of composition of the sample [53-55]. To compensate the Li loss by evaporation, addition of excess Li in the starting material has been often used. However, the excess Li may cause Al contamination due to promotion of Al diffusion from the Al_2O_3 crucible [56]. Although Al contamination is inevitable when Al_2O_3 crucible is used, it may be important for stabilization of the cubic phase [33]. Al works as not only stabilization of cubic phase, but also a sintering additive [32]. Ga also would work similarly to stabilize the cubic phase and increase the Li ion conductivity [38]. Ga must be added intentionally, but Ga exists not only in the crystal of LLZ but also at between the crystal grains. Therefore, precise estimation of amount of substituents and controlling the substitution are very difficult. This can be applied to all substituents. Therefore, lowering the sintering temperature of LLZ has been attempted. Tadanaga et al. used Li_3BO_3 as a sintering additive and sintered LLZ was successfully obtained at 900 °C sintering [57, 58]. This was about 300 °C lower than Murugan's first paper [22]. Also, LLZ is sensitive to sintering atmosphere. The Li ion conductivity of $Li_{6.55}Ga_{0.15}La_3Zr_2O_{12}$ sintered in dry oxygen has higher conductivity than that sintered in air [38, 39, 49, 59]. Li^+/H^+ exchange reaction with adsorbed moisture occurs on the Li-stuffed garnet. As a result, Li ion and moisture form LiOH and then Li_2CO_3 through a reaction with CO_2 in ambient air [60, 61]. The Li^+/H^+ exchange would occur in the tetrahedral sites more than in the octahedral sites [62, 63]. The high sensitivity of LLZ to air, particularly moisture and CO_2 leads to a deviation of composition, resulting in the discrepancy of Li ion conductivity. Additionally, this high sensitivity also causes the complicated tetragonal-cubic phase transition [29, 30, 64-66].

Fabrication of all-solid-state Li battery with LLZ also has been studied. $LiCoO_2$ cathode was prepared by sputtering [32] and sol-gel [67] methods on Al-added LLZ. In the cyclic voltammogram of $LiCoO_2$/Al-added LLZ/Li showed redox peaks of $LiCoO_2$ (Fig. 15). The all-solid-state battery with $LiFePO_4$ [68] and amorphous TiS_4 [69] cathodes were also reported. Ohta et al. prepared $LiCoO_2$ cathode on Nb-substituted LLZ by screen printing and Li metal anode by a vacuum deposition process [70]. The cell demonstrated relatively high performance (discharge capacity was 85 mAh g^{-1} at 0.05 C at 25 °C). Although some groups have succeeded in fabrication of all-solid-state batteries with LLZ electrolyte and Li metal anode, their performance is not sufficient for practical application. High impedance at electrode/electrolyte interface is considered to be a main reason for the low performance. To reduce the impedance at the interface, a composite electrode which comprises active material, carbon (electrical conductive material) and ion conductive ceramics was suggested and fabrication of the all-solid-state battery with the composite electrodes has been studied as well [71].

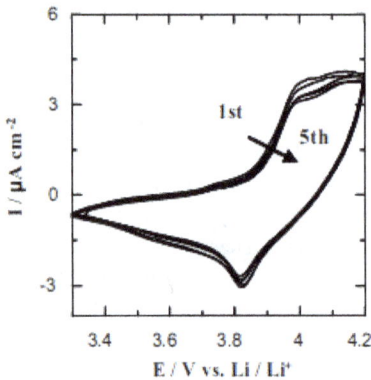

Figure 15. A cyclic voltammogram of Li/Al-added. LLZ/LiCoO₂ cell at room temperature.

4.   Summary

Garnet-type Li ion ceramics are reviewed. Some of the garnet-type ceramics possess extremely high chemical and electrochemical stabilities which allow the fabrication of all-solid-state battery with a Li metal anode. Especially, LLZ has been eagerly researched for all-solid-state battery application due to its high Li ion conductivity. Crystal structure, Li ion conduction path and conductivity have been clarified by many groups. However, a development of LLZ is facing a poor reproducibility. The poor reproducibility is caused by difference of compositions. Many factors influence the composition of LLZ, resulting in the deviation of Li ion conductivity. Establishment of a preparation procedure which is able to control the composition of LLZ precisely is the key issue for further development of LLZ.

A fabrication of all-solid-state battery using LLZ as a solid electrolyte also has been intensively studied. Some groups successfully fabricated all-solid-state batteries. However, their performance is still not sufficient for practical application. The main reason is considered to be high impedance at the electrode/electrolyte interface. Research into new architecture for the interface to reduce the impedance must be done.

References

[1] T. Hasegawa, K. Terabe, T. Sakamoto, M. Aono: Nanoionics switching devices: "Atomic switches", MRS Bull. 34 (2009) 929-934. https://doi.org/10.1557/mrs2009.215

[2] Y. Sun: Lithium ion conducting membranes for lithium-air batteries, Nano Energy 2 (2013) 801-806. https://doi.org/10.1016/j.nanoen.2013.02.003

[3] K. Takada, S. Kondo: Lithium ion conductive glass and its application to solid state batteries, Ionics 4 (1998) 42-47. https://doi.org/10.1007/BF02375778

[4] P. Knauth: Inorganic solid Li ion conductors: an overview, Solid State Ionics 180 (2009) 911-916. https://doi.org/10.1016/j.ssi.2009.03.022

[5] V. Thangadurai, H. Kaack, W. Weppner: Novel fast lithium ion conduction in garnet-type $Li_5La_3M_2O_{12}$ (M=Nb, Ta), J. Am. Ceram. Soc. 86[3] (2003) 437-440. https://doi.org/10.1111/j.1151-2916.2003.tb03318.x

[6] F. Abbattista, M. Vallino and D. Mazza: Remarks on binary system Li_2O-Me_2O_5 (Me = Nb, Ta), Mater. Res. Bull. 22 (1987) 1019-1027. https://doi.org/10.1016/0025-5408(87)90230-3

[7] V. Thangadurai, W. Weppner: Recent progress in solid oxide and lithium ion conducting electrolytes research, Ionics 12 (2006) 81-92. https://doi.org/10.1007/s11581-006-0013-7

[8] E. J. Cussen: The structure of lithium garnets: cation disorder and clustering in a new family of fast Li^+ conductors, Chem. Commun. (2006) 412-413. https://doi.org/10.1039/B514640B

[9] M. Nakayama, M. Kotobuki, H. Munakata, M. Nogami, K. Kanamura: First-principles density functional calculation of electrochemical stability of fast Li ion conducting garnet-type oxides, Phys. Chem. Chem. Phys. 14 (2012) 10008-10014. https://doi.org/10.1039/c2cp40634a

[10] M. Kotobuki, K. Kanamura: Fabrication of all-solid-state battery using $Li_5La_3Ta_2O_{12}$ ceramic electrolyte, Ceramics International 39 (2013) 6481-6487. https://doi.org/10.1016/j.ceramint.2013.01.079

[11] M. Kotobuki, M. Koishi: Preparation of $Li_7La_3Zr_2O_{12}$ solid electrolyte via a sol-gel method, Ceramics International 40 (2014) 5043-5047. https://doi.org/10.1016/j.ceramint.2013.09.009

[12] E. Rossen, J. N. Reimers, J. R. Dahn: Synthesis and electrochemistry of spinel LT-$LiCoO_2$ Solid State Ionics 62 (1993) 53-60. https://doi.org/10.1016/0167-2738(93)90251-W

[13] R. Murugan, W. Weppner, P. Schmid-Beurmann, V. Thangadurai: Structure and lithium ion conductivity of bismuth containing lithium garnets $Li_5La_3Bi_2O_{12}$ and $Li_6SrLa_2Bi_2O_{12}$, Mater. Sci. Eng. B 143 (2007) 14-20. https://doi.org/10.1016/j.mseb.2007.07.009

[14] Y. X. Gao, X. P. Wang, W. G. Wang, Z. Zhuang, D. M. Zhang, Q. F. Fang: Synthesis, ionic conductivity, and chemical compatibility of garnet-like lithium ionic conductor $Li_5La_3Bi_2O_{12}$, Solid State Ionics 181 (2010) 1415-1419. https://doi.org/10.1016/j.ssi.2010.08.012

[15] R. Murugan, W. Weppner, P. Schmid-Beurmann, V. Thangadurai: Structure and lithium ion conductivity of garnet-like $Li_5La_3Sb_2O_{12}$ and $Li_6SrLa_2Sb_2O_{12}$, Mater. Res. Bull. 43 (2008) 2579-2591. https://doi.org/10.1016/j.materresbull.2007.10.035

[16] E. J. Cussen, T. W. S. Yip: A neutron diffraction study of the d^0 and d^{10} lithium garnets $Li_3Nd_3W_2O_{12}$ and $Li_5La_3Sb_2O_{12}$, J. Solid State Chem. 180 (2007) 1832-1839. https://doi.org/10.1016/j.jssc.2007.04.007

[17] V. Thangadurai and W. Weppner: Effect of sintering on the ionic conductivity of garnet-related structure $Li_5La_3Nb_2O_{12}$ and In- and K-doped $Li_5La_3Nb_2O_{12}$, J. Solid State Chem. 179 (2006) 974-984. https://doi.org/10.1016/j.jssc.2005.12.025

[18] I. P. Roof, M. D. Smith, E. J. Cussen, H. C. zur Loye: Crystal growth of a series of lithium garnets $Ln_3Li_5Ta_2O_{12}$ (Ln=La, Pr, Nd): Structural properties, Alexandrite effect and unusual ionic conductivity, J. Solid State Chem. 182 (2009) 295-300. https://doi.org/10.1016/j.jssc.2008.10.032

[19] R. Murugan, V. Thangadurai and W. Weppner: Effect of lithium ion content on the lithium ion conductivity of the garnet-like structure $Li_{5+x}BaLa_2Ta_2O_{11.5+0.5x}$

(x=0-2), Appl. Phys. A: Mater. Sci. Process.91 (2008) 615-620.
https://doi.org/10.1007/s00339-008-4494-2

[20] R. Murugan, V. Thangadurai, W. Weppner: Lattice parameter and sintering
      temperature dependence of bulk and grain-boundary conduction of garnet-like slid
      Li-electrolytes, J. Electrochem. Soc. 155(1) (2008) A90-A101.
      https://doi.org/10.1149/1.2800764

[21] V. Thangadurai, S. Narayanan, D. Pinzaru: Garnet-type solid-state fast Li ion
      conductors for Li batteries: critical review, Chem. Soc. Review 43 (2014) 4714-
      4727. https://doi.org/10.1039/c4cs00020j

[22] R. Murugan, V. Thangadurai, W. Weppner: Fast lithium ion conduction in garnet-
      type $Li_7La_3Zr_2O_{12}$, Angew. Chem.-Int. Ed. 46 (2007) 7778-7781.
      https://doi.org/10.1002/anie.200701144

[23] M. Kotobuki, H. Munakata, K. Kanamura, Y. Sato, T. Yoshida: Compatibility of
      $Li_7La_3Zr_2O_{12}$ solid electrolyte to all-solid-state battery using Li metal anode J.
      Electrochem. Soc. 157[10] (2010) A1076-A1079.
      https://doi.org/10.1149/1.3474232

[24] J. Awaka, A. Takashima, K. Kataoka, N. Kijima, Y. Idemoto, J. Akimoto: Crystal
      structure of fast lithium-ion-conducting cubic $Li_7La_3Zr_2O_{12}$, Chem. Lett. 40 (2011)
      60-62. https://doi.org/10.1246/cl.2011.60

[25] R. Murugan, W. Weppner, P. Schmid-Beurmann, V. Thangadurai: Structure and
      Lithium Ion Conductivity of Bismuth Containing Lithium Garnets $Li_5La_3Bi_2O_{12}$
      and $Li_6SrLa_2Bi_2O_{12}$, Mat. Sci. Eng. B-Solid 143[1-3] (2007) 14-20.
      https://doi.org/10.1016/j.mseb.2007.07.009

[26] Y. Ren, K. Chen, R. Chen, T, Liu, Y. Zhang, C.-W. Nan: Oxide Electrolytes for
      Lithium Batteries, J. Am. Ceram. Soc. 98 (2015) 3603-3623.
      https://doi.org/10.1111/jace.13844

[27] A. Logeat, T. Kohler, U. Eisele, B. Stiaszny, A. Harzer, M. Tovar, A. Senyshyn,
      H. Ehrenberg, B. Kozinsky: From Order to Disorder: The Structure of Lithium
      Conducting Garnets $Li_{7-x}La_3Ta_xZr_{2-x}O_{12}$ (x = 0-2), Solid State Ionics 206[0] (2012)
      33-38. https://doi.org/10.1016/j.ssi.2011.10.023

[28] J. Awaka, N. Kijima, H. Hayakawa, J. Akimoto: Synthesis and structure analysis
      of tetragonal $Li_7La_3Zr_2O_{12}$ with the garnet-related type structure, J. Solid State
      Chem. 182 (2009) 2046-2052. https://doi.org/10.1016/j.jssc.2009.05.020

[29]   G. Larraza, A. Orera, M.L. Sanjuan: Cubic phases of garnet-type $Li_7La_3Zr_2O_{12}$: the role of hydration, J. Mater. Chem. A 1 (2013) 11419-11428. https://doi.org/10.1039/c3ta11996c

[30]   Matsui, K. Sakamoto, K. Takahashi, A. Hirano, Y. Takeda, O. Yamamoto, N. Imanishi: Phase transformation of the garnet structured lithium ion conductor: $Li_7La_3Zr_2O_{12}$, Solid State Ionics 262 (2014) 155-159. https://doi.org/10.1016/j.ssi.2013.09.027

[31]   N. Bernstein, M. D. Johannes, K. Hoang: Origin of the Structural Phase Transition in $Li_7La_3Zr_2O_{12}$, Phys. Rev. Lett. 109[20] (2012) 205702. https://doi.org/10.1103/PhysRevLett.109.205702

[32]   M. Kotobuki, K. Kanamura, Y. Sato, T. Yoshida: Fabrication of all-solid-state lithium battery with lithium metal anode using Al_2O_3-added $Li_7La_3Zr_2O_{12}$ solid electrolyte, J. Power Sources 196 (2011) 7750-7754. https://doi.org/10.1016/j.jpowsour.2011.04.047

[33]   C.A. Geiger, E. Alekseev, B. Lazic, M. Fisch, T. Armbruster, R. Langner, M. Fechtelkord, N. Kim, T. Pettke, W. J. F. Weppner: Crystal Chemistry and Stability of "$Li_7La_3Zr_2O_{12}$" Garnet: A Fast Lithium-Ion Conductor, Inorg. Chem. 50 (2011) 1089-1097. https://doi.org/10.1021/ic101914e

[34]   A. Duvel, A. Kuhn, L. Robben, M. Wilkening, P. Heitjans: Mechanosynthesis of Solid Electrolytes: Preparation, Characterization, and Li Ion Transport Properties of Garnet-Type Al-Doped $Li_7La_3Zr_2O_{12}$ Crystallizing with Cubic Symmetry, J. Phys. Chem. C 116 (2012) 15192-15202. https://doi.org/10.1021/jp301193r

[35]   R. Takano, K. Tadanaga, A. Hayashi, M. Tatsumisago: Low temperature synthesis of Al-doped $Li_7La_3Zr_2O_{12}$ solid electrolyte by a sol-gel process, Solid State Ionics 255 (2014) 104-107. https://doi.org/10.1016/j.ssi.2013.12.006

[36]   J. L. Allen, J. Wolfenstine, E. Rangasamy, J. Sakamoto: Effect of substitution (Ta, Al, Ga) on the conductivity of $Li_7La_3Zr_2O_{12}$, J. Power Sources 206 (2012) 315-319. https://doi.org/10.1016/j.jpowsour.2012.01.131

[37]   Y. Li, C.-A. Wang, H. Xie, J. Cheng, J.B. Goodenough: High lithium ion conduction in garnet-type $Li_6La_3ZrTaO_{12}$, Electrochem. Commun. 13 (2011) 1289-1292. https://doi.org/10.1016/j.elecom.2011.07.008

[38]   H. El Shinawi, J. Janek: Stabilization of cubic lithium-stuffed garnets of type "$Li_7La_3Zr_2O_{12}$" by addition of gallium, J. Power Sources 225 (2013) 13-19. https://doi.org/10.1016/j.jpowsour.2012.09.111

[39]   M.A. Howard, O. Clemens, E. Kendrick, K.S. Knight, D.C. Apperley, P.A.
        Anderson, P.R. Slater: Effect of Ga incorporation on the structure and Li ion
        conductivity of $La_3Zr_2Li_7O_{12}$, Dalton Trans. 41 (2012) 12048-12053.
        https://doi.org/10.1039/c2dt31318a

[40]   M. Huang, A. Dumon, C.-W. Nan: Effect of Si, In and Ge Doping on High Ionic
        Conductivity of $Li_7La_3Zr_2O_{12}$, Electrochem. Commun. 21[0] (2012) 62-64.
        https://doi.org/10.1016/j.elecom.2012.04.032

[41]   S. Ohta, T. Kobayashi, T. Asaoka: High Lithium Ionic Conductivity in the Garnet-
        Type Oxide $Li_{7-x}La_3(Zr_{2-x}, Nb_x)O_{12}$ (X = 0-2), J. Power Sources 196[6] (2011)
        3342-3345. https://doi.org/10.1016/j.jpowsour.2010.11.089

[42]   C. Deviannapoorani, L. Dhivya, S. Ramakumar, R. Murugan: Lithium ion
        Transport Properties of High Conductive Tellurium Substituted $Li_7La_3Zr_2O_{12}$
        Cubic Lithium Garnets, J. Power Sources 240[0] (2013) 18-25.
        https://doi.org/10.1016/j.jpowsour.2013.03.166

[43]   R. Murugan, S. Ramakumar, and N. Janani: High Conductive Yttrium Doped
        $Li_7La_3Zr_2O_{12}$ Cubic Lithium Garnet, Electrochem. Commun. 13[12] (2011) 1373-
        1375. https://doi.org/10.1016/j.elecom.2011.08.014

[44]   S. Song, B. Yan, F. Zheng, H. M. Duong, L. Lu: Crystal Structure, Migration
        Mechanism and Electrochemical Performance of Cr-Stabilized Garnet, Solid State
        Ionics 268 (2014) 135-139. https://doi.org/10.1016/j.ssi.2014.10.009

[45]   M. Matsui, K. Takahashi, K. Sakamoto, A. Hirano, Y. Takeda, O. Yamamoto, N.
        Imanishi: Phase stability of a granet-type lithium ion conductor $Li_7La_3Zr_2O_{12}$,
        Dalton Trans. 43 (2014) 1019-1024. https://doi.org/10.1039/C3DT52024B

[46]   H. Xie, J. A. Alonso, Y. Li, M. T. Fern andez-Diaz, J. B. Goodenough: Lithium
        Distribution in Aluminum-Free Cubic $Li_7La_3Zr_2O_{12}$, Chem. Mater. 23[16] (2011)
        3587-3589. https://doi.org/10.1021/cm201671k

[47]   Y. Kihira, S. Ohta, H. Imagawa, T. Asaoka: Effect of Simultaneous Substitution of
        Alkali Earth Metals and Nb in $Li_7La_3Zr_2O_{12}$ on Lithium-Ion Conductivity, ECS
        Electrochem. Lett. 2[7] (2013) A56-A59. https://doi.org/10.1149/2.001307eel

[48]   S.-W. Baek, J.-M. Lee, Y. Kim, M.-S. Song, Y. Park: Garnet related lithium ion
        conductor processed by spark plasma sintering for all solid state batteries, J. Power
        Sources 249 (2014) 197-206. https://doi.org/10.1016/j.jpowsour.2013.10.089

[49]   C. Bernuy-Lopez, W. Manalastas Jr., J. M. Lopez del Amo, A. Aguadero, F.
        Aguesse, J. A. Kilner: Atmosphere Controlled Processing of Ga-Substituted

Garnets for High Li-Ion Conductivity Ceramics, Chem. Mater. 26[12] (2014) 3610-3617. https://doi.org/10.1021/cm5008069

[50] Y. Li, J.-T. Han, C.-A. Wang, H. Xie, J. B. Goodenough: Optimizing Li^+ Conductivity in a Garnet Framework, J. Mater. Chem. 22[30] (2012) 15357-15361. https://doi.org/10.1039/c2jm31413d

[51] Y. Wang, W. Lai: High Ionic Conductivity Lithium Garnet Oxides of $Li_{7-x}La_3Zr_{2-x}Ta_xO_{12}$ Compositions, Electrochem. Solid-State Lett. 15[5] (2012) A68-A71. https://doi.org/10.1149/2.024205esl

[52] D. Wang, G. Zhong, O. Dolotko, Y. Li, M. J. McDonald, J. Mi, R. Fu, Y. Yang: The Synergistic Effects of Al and Te on the Structure and Li^+-Mobility of Garnet-Type Solid Electrolytes, J. Mater. Chem. A 2[47] (2014) 20271-20279. https://doi.org/10.1039/C4TA03591G

[53] Y. Ren, H. Deng, R. Chen, Y. Shen, Y. Lin, C.-W. Nan: Effects of Li Source on Microstructure and Ionic Conductivity of Al-Contained $Li_{6.75}La_3Zr_{1.75}Ta_{0.25}O_{12}$ Ceramics, J. Eur. Ceram. Soc. 35[2] (2015) 561-572. https://doi.org/10.1016/j.jeurceramsoc.2014.09.007

[54] M. Huang, T. Liu, Y. Deng, H. Geng, Y. Shen, Y. Lin, C.-W. Nan: Effect of Sintering Temperature on Structure and Ionic Conductivity of $Li_{7-x}La_3Zr_2O_{12-0.5x}$(x = 0.5-0.7) Ceramics, Solid State Ionics 204-205[0] (2011) 41-45. https://doi.org/10.1016/j.ssi.2011.10.003

[55] R.-J. Chen, M. Huang, W.-Z. Huang, Y. Shen, Y.-H. Lin, C.-W. Nan: Effect of Calcining and Al Doping on Structure and Conductivity of $Li_7La_3Zr_2O_{12}$, Solid State Ionics 265[0] (2014) 7-12. https://doi.org/10.1016/j.ssi.2014.07.004

[56] K. Liu, J.-T. Ma, C.-A. Wang: Excess Lithium Salt Functions More Than Compensating for Lithium Loss When Synthesizing $Li_{6.5}La_3Ta_{0.5}Zr_{1.5}O_{12}$ in Alumina Crucible, J. Power Sources 260[0] (2014) 109-114. https://doi.org/10.1016/j.jpowsour.2014.02.065

[57] K. Tadanaga, R. Takano, T. Ichinose, S. Mori, A. Hayashi, M. Tatsumisago: Low temperature synthesis of highly ion conductive $Li_7La_3Zr_2O_{12}$-Li_3BO_3composites, Electrochem. Comm. 33 (2013) 51-54. https://doi.org/10.1016/j.elecom.2013.04.004

[58] R. Takano, K. Tadanaga, A. Hayashi, M. Tatsumisago: Low temperature synthesis of Al-doped $Li_7La_3Zr_2O_{12}$ solid electrolyte by a sol-gel process, Solid State Ionics 255 (2014) 104-107. https://doi.org/10.1016/j.ssi.2013.12.006

[59]   J. Wolfenstine, J. Ratchford, E. Rangasamy, J. Sakamoto, J. L. Allen: Synthesis and High Li-Ion Conductivity of Ga-Stabilized Cubic $Li_7La_3Zr_2O_{12}$, Mater. Chem. Phys. 134[2-3] (2012) 571-575.
https://doi.org/10.1016/j.matchemphys.2012.03.054

[60]   X.-P. W. W.-G. Wang, Y.-X. Gao, J.-F. Yang, Q.-F. Fang: Investigation on the Stability of $Li_5La_3Ta_2O_{12}$ Lithium Ionic Conductors in Humid Environment, Front. Mater. Sci. 4[2] (2010) 189-192. https://doi.org/10.1007/s11706-010-0017-0

[61]   C. Galven, J.-L. Fourquet, M.-P. Crosnier-Lopez, F. Le Berre: Instability of the Lithium Garnet $Li_7La_3Sn_2O_{12}$: Li^+/H^+ Exchange and Structural Study, Chem. Mater. 23[7] (2011) 1892-1900. https://doi.org/10.1021/cm103595x

[62]   C. Li-quan, W. Lian-zhong, C. Guang-can, W. Gang, L. Z-rong: Investigation of New Lithium Ionic Conductors $Li_{3+x}V_{1-x}Si_xO_4$, Solid State Ionics 9-10 (1983) 149-152. https://doi.org/10.1016/0167-2738(83)90224-2

[63]   A. Khorassani, A. R. West: Li^+ Ion Conductivity in the System Li_4SiO_4-Li_3VO_4, J. Solid State Chem. 53[3] (1984) 369-375. https://doi.org/10.1016/0022-4596(84)90114-2

[64]   S. Toda, K. Ishiguro, Y. Shimonishi, A. Hirano, Y. Takeda, O. Yamamoto, N. Imanishi: Low Temperature Cubic Garnet-Type CO_2-Doped $Li_7La_3Zr_2O_{12}$, Solid State Ionics 233[0] (2013) 102-106. https://doi.org/10.1016/j.ssi.2012.12.007

[65]   X. P. Wang, Y. Xia, J. Hu, Y. P. Xia, Z. Zhuang, L. J. Guo, H. Lu, T. Zhang, Q. F. Fang: Phase Transition and Conductivity Improvement of Tetragonal Fast Lithium Ionic Electrolyte $Li_7La_3Zr_2O_{12}$, Solid State Ionics 253[0] (2013) 137-142. https://doi.org/10.1016/j.ssi.2013.09.029

[66]   Y. Wang, W. Lai: Phase Transition in Lithium Garnet Oxide Ionic Conductors $Li_7La_3Zr_2O_{12}$: The Role of Ta Substitution and H_2O/CO_2 Exposure, J. Power Sources 275[0] (2015) 612-620. https://doi.org/10.1016/j.jpowsour.2014.11.062

[67]   M. Kotobuki, K. Kanamura, Y. Sato, K. Yamamot, T. Yoshida: Electrochemical properties of $Li_7La_3Zr_2O_{12}$ solid electrolyte prepared in argon atmosphere, J. Power Sources 199 (2012) 346-349.
https://doi.org/10.1016/j.jpowsour.2011.10.060

[68]   C.-W. Ahn, J.-J. Choi, J. Ryu, B.-D. Hahn, J.-W. Kim, W.-H. Yoon, J.-H. Choi, J.-S. Lee, D.-S. Park: Electrochemical properties of $Li_7La_3Zr_2O_{12}$-based solid state battery, J. Power Sources 272 (2014) 554-558.
https://doi.org/10.1016/j.jpowsour.2014.08.110

[69]   T. Matsuyama, R. Takano, K. Tadanaga, A. Hayashi, M. Tatsumisago: Fabrication of all-solid-state lithium secondary batteries with amorphous TiS_4 positive electrodes and $Li_7La_3Zr_2O_{12}$ solid electrolytes, Solid State Ionics 285 (2016) 122-125. https://doi.org/10.1016/j.ssi.2015.05.025

[70]   S. Ohta, S. Komagata, J. Seki, T. Saeki, S. Morishita, T. Asaoka: All-solid-state lithium ion battery using garnet-type oxide and Li_3BO_3 solid electrolytes fabricated by screen-printing, J. Power Sources 238 (2013) 53-56. https://doi.org/10.1016/j.jpowsour.2013.02.073

[71]   A. Aboulaich, R. Bouchet, G. Delaizir, V. Seznec, L. Tortet, M. Morcrette, P. Rozier, J.-M. Tarascon, V. Viallet, M. Dolle: A New Approach to Develop Safe All-Inorganic Monolithic Li-Ion Batteries, Adv. Energy Mater. 1 [2] (2011) 179-183. https://doi.org/10.1002/aenm.201000050

CHAPTER 8

Redox Flow Lithium Batteries

Feng Pan, Yun Guang Zhu and Qing Wang

Department of Materials Science and Engineering, National University of Singapore, Singapore

qing.wang@nus.edu.sg

Abstract

The penetration of battery technologies has been expanding from portable electronics to electric vehicles and power grids. The increasing demand for energy storage in new areas drives the development of battery technologies with longer lifetime, better safety and lower cost. Redox flow batteries as a distinct energy storage system from those with enclosed configurations, present decoupled energy storage and power generation and have been taken as an important candidate for large-scale stationary energy storage. In this chapter, the development of redox flow batteries is briefly reviewed and the issues confronted by the conventional flow batteries are analyzed. As an implementable solution, redox flow lithium batteries based on redox targeting reactions between redox mediators and battery materials are introduced. Its working principle, research progress and future development are discussed in detail.

Keywords

Redox Flow Battery, Redox Targeting, Energy Density, Redox Flow Lithium Battery

Contents

1.   Conventional Redox Flow Batteries

Large-scale stationary energy storage is becoming increasingly important for power grid, backup power and instantaneous storage of renewable energy sources (i.e. solar, wind, etc.)[1]. Among various electrochemical energy storage technologies, the redox flow battery (RFB) is considered suitable for large-scale deployment due to its modular design, good scalability and operation flexibility. As Figure 1 shows, a typical RFB consists of three parts: stack cell, storage tanks and flow system [2]. Redox species are dissolved in liquid electrolytes (catholyte and anolyte), which are stored in the tanks, circulated through the system and react in the cell upon operation. As such, the energy storage and power generation are decoupled: the size of the tanks determines the capacity and energy stored, while the area of stack cell determines the power, which makes it possible for independent control of overall capacity and power to meet various requirements [3].

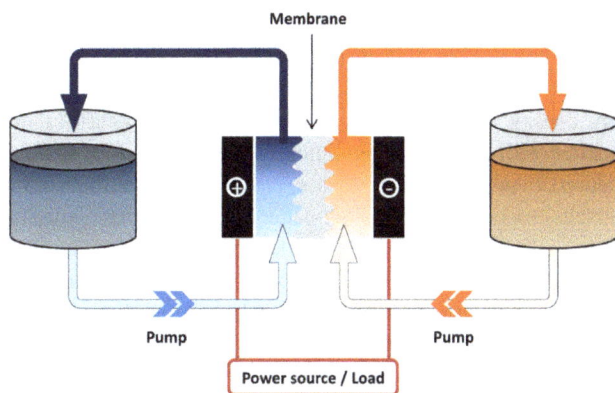

Figure 1. Schematic illustration of a redox flow battery. The active species are stored externally in two storage tanks, while the conversion between electrical and chemical energy occurs in the stack cell.

RFBs can be classified as aqueous and non-aqueous systems in terms of the electrolyte solvent. The first RFB is the aqueous Fe-Cr system primarily developed at NASA in the 1970s [4], which employs Fe^{2+}/Fe^{3+} and Cr^{2+}/Cr^{3+} redox couples as the positive and negative redox species, respectively, and delivers a cell voltage of 1.18 V. The Fe-Cr flow battery suffers from problems, such as the sluggish kinetics of Cr^{2+}/Cr^{3+} and H_2 evolution, which constraints its further development. A range of redox species has been extensively studied to replace the Fe-Cr system, such as all-vanadium [5], vanadium-bromine [6], polysulfide-bromine [7], zinc-bromine [8], etc.

1.1 All Vanadium Redox Flow Battery

Among various RFB systems, the all vanadium redox-flow battery (VRB) is the most developed one. VRB employs V^{2+}/V^{3+} as the redox species in the anolyte and VO^{2+}/VO_2^{+} in the catholyte, rendering a cell voltage of about 1.26 V:

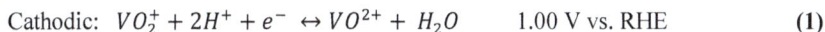

Cathodic: $VO_2^{+} + 2H^{+} + e^{-} \leftrightarrow VO^{2+} + H_2O$ 1.00 V vs. RHE   **(1)**

Anodic: $V^{2+} - e^{-} \leftrightarrow V^{3+}$ -0.26 V vs. RHE   **(2)**

As both the catholyte and anolyte share the same vanadium element, electrolyte contamination caused by crossover is greatly alleviated, which allows for a long lifetime of about 15-20 years [9]. Since the report by Skyllas-Kazacos and co-workers in the 1980s, extensive studies have been carried out on VRB, and demonstration systems as large as MW/MWh have been installed [3] [10-20]. At present, the obstacle for its commercialization is the high cost, which is partly associated with the low energy density. In a typical VRB, the concentration of redox species is in the range of 1.5-3.0 M, and the energy density is 20-30 Wh/L. Such a low energy density makes VRB less attractive compared with rivals like the lithium ion battery. Various strategies have been proposed to address these issues. For instance, researchers at Pacific Northwest National Laboratory (PNNL) reported a Fe-V flow battery, combining the advantages of Fe-Cr system (low cost) and VRB (relatively good kinetics) while avoiding their drawbacks [21]. In addition, researchers at the University of Twente proposed a vanadium-air flow battery, in which the anolyte contains V^{2+}/V^{3+} and O_2 is fed from air into the cathodic side [22-23]. Since the catholyte is replaced by an air electrode, the vanadium-air battery has the potential to achieve higher energy density.

1.2 Zn-Br Battery

The Zn-Br battery (ZBB) utilizes Br^-/Br_2 and Zn/Zn^{2+} as the redox species in catholyte and anolyte, respectively [24]. Since the Zn electrode is a non-liquid reactant, ZBB is considered as a hybrid flow battery. The reactions of ZBB battery are shown below:

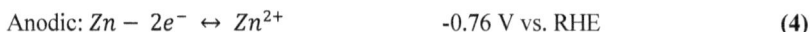

Cathodic: $Br_2 + 2e^- \leftrightarrow 2Br^-$ 1.09 V vs. RHE   **(3)**

Anodic: $Zn - 2e^- \leftrightarrow Zn^{2+}$ -0.76 V vs. RHE   **(4)**

The synergy between the dissolved Br^-/Br_2 in the catholyte and Zn anode brings about significant enhancement in energy density since the volume of anolyte is greatly reduced. However, ZBB battery suffers from the poor miscibility and corrosive nature of Br_2. A number of redox species have been proposed to replace the bromine catholyte [25], such as Zn-Ce battery [26], Zn-Fe battery [27], and Zn-polyiodide battery [26-28], etc., while limited success has been achieved so far. In addition, the cycling stability of Zn anode presents another challenge that needs to be addressed.

1.3 Organic-inorganic Aqueous Flow Battery

Aziz group developed a metal-free flow battery based on 9,10-anthraquinone-2,7-disulphonic acid (AQDS) [29]. In their work, the rapid and reversible two-electron two-proton reduction of AQDS was exhibited in sulfuric acid. Combined with Br^-/Br_2, the power of the RFB could reach 0.60 W/cm^2 at 1.30 A/cm^2. Later, the group reported a more stable and safer aqueous RFB system based on 2,6-dihydroxyanthraquinone (DHAQ) and ferricyanides in alkaline solutions [30]. Recently, the same group utilized a radical-free organic redox compound, alloxazine, a tautomer of the isoalloxazine backbone of vitamin B_2, in aqueous RFBs [31], which exhibits an open-circuit voltage

(OCV) of 1.20 V, excellent coulombic efficiency (99.7% per cycle) and capacity retention (99.98% per cycle). These organic-inorganic aqueous flow battery systems exclude the use of expensive chemicals and potentially enable cost-effective large-scale energy storage as compared to the VRBs.

1.4   Non-aqueous Systems

The electrochemical window of water is 1.23 V, which constraints the cell voltage of aqueous RFBs. As such, non-aqueous electrolytes are advantageous since they usually have a much wider electrochemical window, which meanwhile allows for a greater variety of redox species. A large number of redox species have been reported for non-aqueous redox flow batteries, and most of them can be classified into two categories: metal complexes and metal-free compounds. Among the metal complexes, vanadium acetylacetonate (V(acac)$_3$) has attracted most attention. Cyclic voltammetric test of V(acac)$_3$ shows two redox peaks at -1.75 V and 0.45 V vs. Ag/Ag$^+$ in various organic electrolytes. Thus it can be used in both the anolyte and catholyte, delivering a cell voltage of ~2.20 V [32]. The solubility of V(acac)$_3$ ranges from 0.50 to 0.80 M in different solvents [33]. However, only maximally 0.05 M V(acac)$_3$ was demonstrated in the charge/discharge tests, making the capacity far too low for real application. In addition, the system is vulnerable to moisture, which requires proper sealing of the RFB system for good cycling stability.

1.5   Hybrid Lithium Redox Flow Battery

Lithium metal has a high specific capacity of 3860 mAh/g. Meanwhile, as an anode material, the redox potential of Li is as low as -3.04 V vs. SHE, which is advantageous to obtain high cell voltage when combined with a cathode [34]. Therefore, Li metal has been extensively employed as the anode material to improve the energy density of RFBs. As shown in Figure 2, the cell stack of a hybrid lithium RFB consists of three parts: a lithium metal anode, a flow-through cathode connected to a storage tank, and a solid electrolyte membrane [35]. The electrode reactions of hybrid lithium RFB can be described as follows:

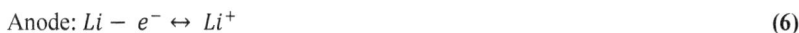

Cathode: $Z^+ + e^- \leftrightarrow Z$ (5)

Anode: $Li - e^- \leftrightarrow Li^+$ (6)

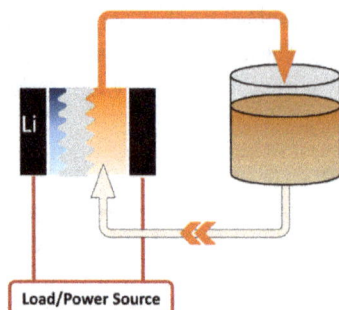

Figure 2. Schematic illustration of a lithium hybrid flow battery. The anodic compartment consists of lithium metal as anode, while the cathodic compartment contains catholyte circulated between the electrode and a storage tank.

Where Z/Z^+ is the catholyte redox species. For hybrid lithium RFB, as the anodic potential and capacity is fixed by lithium metal, the selection of catholyte is critical for the overall performance of the system. In order to achieve high energy density, catholyte is desired to have a high concentration and positive redox potential. Numerous redox mediators have been studied for the catholyte, ranging from aqueous system to non-aqueous system, including transition metal ions, halogen ions [36-39], metal complexes [40], sulfur/polysulfide [41], metal-free species [42], etc. However, as most of the other non-aqueous RFBs, the system severely suffers from low Li^+ conductivity through the membrane. In addition, the stability of lithium metal upon prolonged cycling is yet satisfactory. These represent the major obstacles for the further development of hybrid lithium RFBs.

1.6   Semi-solid Flow Battery

The concentration of redox species in a solid material is generally much higher than its liquid counterpart. For instance, the theoretical molar concentration of Co^{2+}/Co^{3+} in $LiCoO_2$ is about 51.63 M (equivalent to Li^+ concentration in the same material), which is hardly achieved in a liquid solution. In a semi-solid flow battery, solid active materials suspension is used to replace the dissolved redox species in electrolyte (Figure 3), so the energy density is no longer constrained by the solubility of redox species [43]. Duduta *et al.* reported a semi-solid lithium rechargeable flow battery, in which $LiCoO_2$ slurry (20 *vol.*%, 10.2 M, 1.5% Ketjen black) was employed as the catholyte material and $Li_4Ti_5O_{12}$ slurry (10 *vol.*%, 2.3 M, 2% Ketjen black) as the anolyte. The equivalent concentration of the catholyte is 5 times higher than the VRBs. Another advantage is that since there isn't soluble redox mediator in the electrolyte, a low-cost porous separator could be used to replace the ion-exchange membrane. Besides the commonly studied Li^+ ion storage materials, other types of semi-solid flow batteries, such as Li-sulfur flow battery were also reported [44-45]. In addition, Lu *and co-workers* demonstrated a multiple redox semi-solid-liquid flow battery [45], where lithium iodide liquid catholyte was paired with semi-solid sulphur/carbon composite in the flow cell, achieving a high volumetric capacity (550 Ah/L in the catholyte).

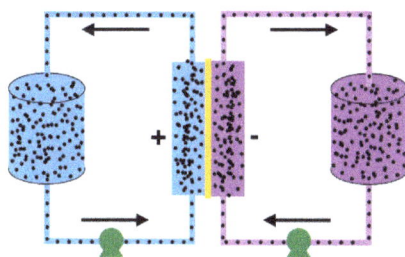

Figure 3. Diagram showing the configuration of a semi-solid flow battery. Both anolyte and catholyte contain suspensions of active solid particles.

While promising, a critical issue for the semi-solid flow battery is that although carbon additives are usually added in the active material slurries, the electronic conductivity of the suspension is still much lower than its ionic conductivity. Another major concern is the high viscosity of the fluids that ensues from the increasing loading of solid material, making the fluids management difficult. This problem could partly be alleviated by tailoring the interactions between suspension particles, but the active materials fraction is still limited to around 20 *vol.*% in most reports [46]. Given these limitations, the high capacity of solid active materials may not be fully exploited in the semi-solid configuration.

2. Redox Flow Lithium-ion Battery

2.1 Redox Targeting Reactions of Battery Materials

The redox targeting concept was first proposed by Wang *et al.* in 2006 [47]. In the presence of redox shuttle molecules, an active electrode material can be charged/discharged *via* redox targeting reactions without being attached to the current collector. The exchange of electrons between the active material and the electrode is mediated by the diffusion of redox shuttle molecules in the electrolyte, which allows the solid active materials to be charged/discharged without conducting additives (Figure 4) [47]. The application of redox targeting reactions to both the cathode and anode leads to a novel energy storage device —*Redox-Flow Lithium Battery* (RFLB). In 2013, Huang *et al.* reported a RFLB half-cell based on the redox targeting reactions between redox mediators and a cathodic battery material [48]. Unlike the semi-solid configuration, RFLB holds the solid materials statically in the tank, while the redox shuttle molecules circulate in the flow system. The electron transfer between the solid materials and current connector is conducted *via* the circulation of shuttle molecules.

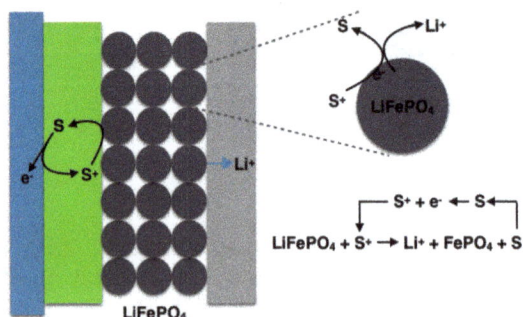

Figure 4. The principle of redox targeting of an insulating electrode material such as LiFePO$_4$ by a freely diffusing molecular shuttle S. (Figure reproduced according to reference 47)

2.2   Redox Flow Lithium-ion Battery Half Cell

In the reported RFLB half-cell, lithium metal was used as anode, dibromoferronce (FcBr$_2$) and ferrocene (Fc) were selected as the oxidation and reduction agents in the catholyte, and LiFePO$_4$ powder was placed in a separate storage tank connected with the cathodic compartment [48]. LiFePO$_4$ typically shows a potential of 3.45 V vs. Li/Li$^+$, while FcBr$_2$ is at 3.65 V and Fc at 3.25 V. Hence LiFePO$_4$ can be chemically delithiated (oxidized) and lithiated (reduced) by these two molecules, respectively. As a result, the charge and discharge process could be achieved by the redox targeting reactions between LiFePO$_4$ and shuttle molecules. Because the solid materials are immobile, the viscosity of catholyte is not a concern in RFLB. Meanwhile, no conducting additive is required in the charge/discharge process, with which a large fraction of solid active materials is allowed in the storage tank. The lithium concentration in $LiFePO_4$ is 22.8 M — when filled in the tank it would be well above 10 M for a porosity of 50% (Note: the porosity here and in the following sections is just a number that has been tested for comparison, not an optimized value) for electrolyte flowing through. In addition, since most of the energy is stored in the solid materials in the tank, the requirement for high-concentration redox shuttle molecules becomes less, which is however challenging for most non-aqueous electrolytes. Lower electrolyte concentration is also beneficial to the reduction of crossover contamination.

On the other hand, Pan *et al.* reported anatase TiO_2 (~1.80 V vs. Li/Li^+) chemically lithiated/delithiated by bis(pentamethylcyclopentadienyl)cobalt ($CoCp^*_2$, 1.36 V vs. Li/Li^+) and cobaltocene (1.95 V vs. Li/Li^+) as the anodic Li^+ storage material [49]. To improve the voltage efficiency, bis(pentamethylcyclopentadienyl)chromium ($CrCp^*_2$, 1.77 V vs. Li/Li^+) was further studied to replace $CoCp^*_2$ for the lithiation of TiO_2 [50]. In order to simplify the system, Huang *et al.* developed a RFLB half-cell based on a single redox mediator (LiI) for the reversible delithiation/lithiation of $LiFePO_4$ [51]. Upon charging, I^- is oxidized to I_3^- and higher order polyiodides, which is able to oxidize $LiFePO_4$ (delithiation), while upon discharging, the formed I^- reduces $FePO_4$ back to $LiFePO_4$ (lithiation). Due to the good stability of LiI, the LiI-$LiFePO_4$ RFLB half-cell presents good cycling performance and capacity retention. The theoretical energy density of the cell can reach as high as 670 Wh/L when the tank is filled with $LiFePO_4$ with a porosity of 50%. Besides, sulfur was also employed in RFLB half-cell as a cathodic active material in the tank, which in theory could further boost the energy density [52].

2.3   Redox Flow Lithium-ion Battery Full Cell

As other lithium hybrid RFBs, the cycling performance of RFLB half-cells is badly restricted by the stability of Li metal. Therefore, the development of RFLB full cell by replacing the Li anode with a flowable redox system is desired. While such a development seems intuitively natural by simply combining two RFLB half-cells, the RFLB full cell has a higher demand for robust Li^+ ion conducting membrane to prevent the crossover of redox mediators. Based on the redox targeting reactions of anodic material (TiO_2) and cathodic material ($LiFePO_4$), Jia *et al.* reported a RFLB full cell, in which a Nafion-PVDF composite membrane was developed to segregate the anodic and cathodic compartment while allows for the transport of Li^+, as shown in Figure 5 [53]. The tank energy density of the RFLB full cell is ~500 Wh/L considering a 50% porosity of solid active materials in the tank, which is 10 times higher than that of a VRB. While a functional RFLB full cell has been developed, there are a few issues to be addressed: (1) Li^+ conductivity of membrane, which remains a limiting factor for the power performance of the system; (2) redox mediators with suitable redox potential and chemical robustness, for reduced voltage loss and enhanced stability; (3) microstructures, porosity, and tortuosity of solid active materials in the tank, to facilitate the redox targeting reactions and enable improved energy density.

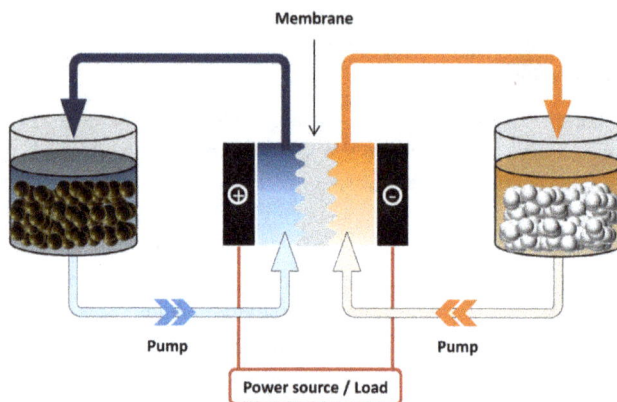

Figure 5. Schematic illustration of a RFLB full cell. The solid active materials are stored statically in the tanks, while redox molecules dissolved in the electrolytes are circulated between the respective storage tank and electrode.

3.   Redox Flow Lithium Oxygen Battery

3.1   Redox Mediators for Lithium Oxygen Batteries

Lithium-oxygen (Li-O$_2$) battery, due to its high-energy density, is believed to be one of the most promising energy storage technologies in the future. However, the development of a Li-O$_2$ battery is presently hindered by large overpotentials, poor cycling stability (degradation of electrolytes and cathodes, lithium metal), etc. This is particularly severe for aprotic systems as the discharge product Li$_2$O$_2$ is insulating and insoluble which passivates and clogs the cathode for further reaction. While numerous heterogeneous catalysts have been designed to lower the overpotentials and enhance the cycling performance of Li-O$_2$ batteries, most of the problems remain. Recently, soluble redox catalysts have been introduced for aprotic Li-O$_2$ batteries to alleviate the passivation and clogging issues [54-56].

Figure 6a shows the redox-mediated oxygen reduction reaction (ORR) in an aprotic Li-O$_2$ battery, for which the redox mediator is electrochemically reduced on the electrode, which then chemically reduces O$_2$ to form Li$_2$O$_2$ in the electrolyte. The redox-mediated oxygen evolution reaction (OER) is shown in Figure 6b — the OER redox mediator is firstly oxidized on the electrode, which then oxidizes Li$_2$O$_2$ to evolve O$_2$. As a thermodynamic requirement, the redox potential of the ORR mediator should be more negative than that of the ORR reaction, while the redox potential of the OER mediator should be more positive than that of the OER reaction. A number of redox mediators have been attempted to lower the overpotentials and improve the round-trip energy efficiency of Li-O$_2$ batteries [54-56]. However, these issues are unlikely to be thoroughly addressed considering the formed Li$_2$O$_2$ remains in the cathode with the conventional cell configuration.

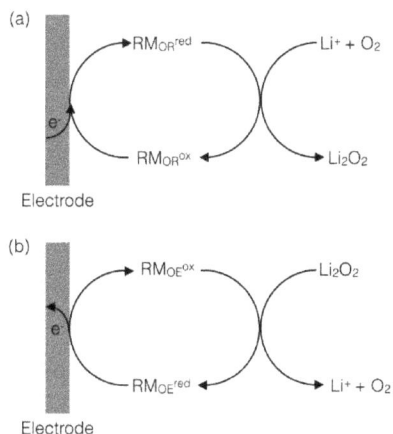

Figure 6. Schematic showing the principles of redox mediated (a) ORR and (b) OER reactions on the cathode of aprotic Li-O$_2$ batteries.

When an aprotic Li-O$_2$ battery is fully discharged, the pores of cathode will eventually be filled with Li$_2$O$_2$. Hence the capacity of Li-O$_2$ batteries is limited by the effective volume of the cathodic compartment. When only OER redox mediators are employed, the ORR reaction takes place directly at the cathode/electrolyte interface and Li$_2$O$_2$ can only be deposited on the electrode through electrochemical reactions (Figure 7) [57]. In such a case, passivation of the cathode would be inevitable, which also hinders the ORR reaction of redox mediators in the subsequent charging process. When only ORR redox mediators are introduced into the Li-O$_2$ battery, the formation of Li$_2$O$_2$ occurs not only on the electrode surface but also in the pores due to the homogeneous redox targeting reactions in the solution phase, which results in an enhanced discharge capacity [58]. However, this may impair the electrochemical oxidation of Li$_2$O$_2$ during the charging process since the contact of the chemically formed Li$_2$O$_2$ with electrode is less favorable. When both the OER and ORR redox catalysts are used in Li-O$_2$ batteries, it is expected that both the discharge capacity and recharge ability would be improved provided that both the redox catalysts survive the aggressive conditions of the battery.[59]

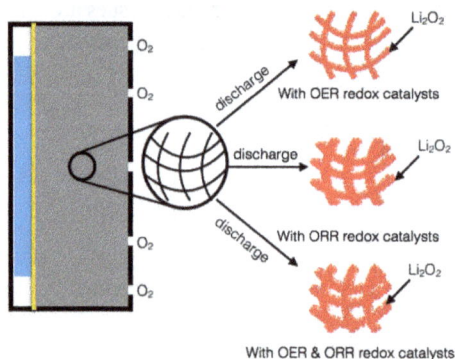

Figure 7. Deposition of Li$_2$O$_2$ in the cathode in the presence of different redox catalysts.

3.2   Redox Flow Lithium Oxygen Batteries

Combining a flow battery with the lithium-oxygen system is deemed to provide a new approach to enhance the energy density of flow batteries. For instance, Zhou *and co-workers* demonstrated a Li-air fuel cell system with a circulating catholyte [60-61], in which the solid LiOH and water are separated from the catholyte and the battery could be "chemically recharged" outside of the cathode compartment [62]. By integrating the redox-mediated ORR and OER reactions into a flow system, Wang group firstly demonstrated a rechargeable redox flow lithium-oxygen battery (RFLOB), which elegantly addresses the passivation and clogging issues confronted by conventional Li-O$_2$ batteries [63-64].

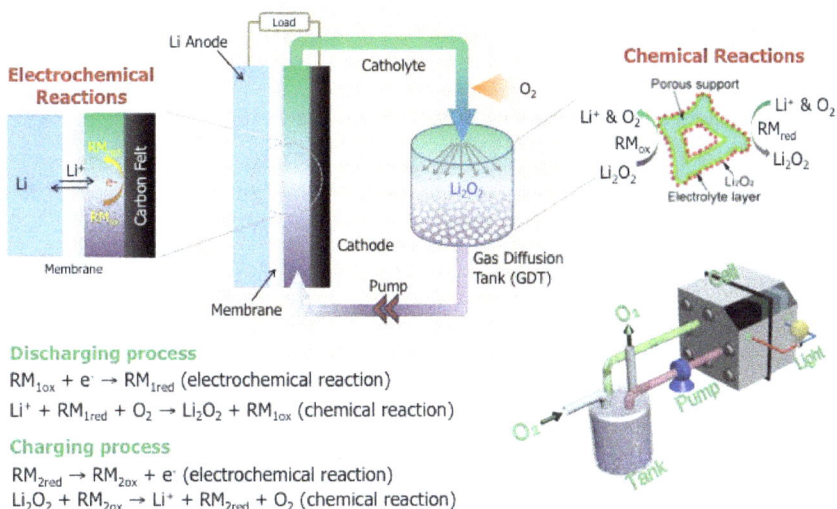

Figure 8. Schematic illustrating the working principle of RFLOB. The structure of RFLOB and the reactions involved in the charging and discharging process are shown. (The figure is re-drawn according to reference 63 and 64.)

As shown in Figure 8, RFLOB consists of a cell stack and a separated gas diffusion tank (GDT) connected to the cathodic compartment of the cell. To prevent the Li anode from parasitic reactions with oxygen and the redox mediators, a Li^+-conducting membrane is used between the anodic and cathodic compartments. During the discharging process, the ORR redox catalyst (i.e. ethyl viologen or benzoquinone derivatives [63-64]) is firstly reduced in the cell and then flows into the GDT, where the oxygen is reduced to form Li_2O_2 by the reduced ORR redox catalyst. Apparently, the passivation and clogging of the cathode are obviated as the formation of Li_2O_2 takes place in the GDT. During the charging process, OER redox catalyst (i.e. iodine or triarylamine derivatives [63-64]) is firstly oxidized on the cathode and flows into the GDT, where Li_2O_2 is decomposed into Li^+ and O_2. The detailed reactions involved in the discharging and charging processes are shown in Figure 8. The spatial separation of the electrochemical reactions (on the cathode) and the chemical reactions (in the GDT) brings about important advantages for the long-term operation of Li-O_2 batteries. Besides the obviation of surface passivation and pore clogging of the cathode, as oxygen is not fed into the cell, the Li anode is further protected from attacks of species from air. In addition, with such a configuration the capacity of RFLOB would not be limited by the volume of cathodic compartment. So the discharge depth could be greatly boosted.

4.   Challenges and Opportunities of Redox Flow Batteries

The past 5 years has witnessed a surge of studies on RFBs and a variety of RFB systems have been developed. While these new systems have shown advancement in terms of one or several performance indicators, there are still many challenges to overcome before they are developed into commercially viable systems for practical applications. The conventional systems, including the most studied VRB systems with soluble redox species in both the anolyte and catholyte, are restricted by the solubility of the redox species, which limits the reachable energy density and impacts the cost. While the hybrid RFBs using Zn or Li metal anode demonstrated significantly improved energy density, their operation flexibility is compromised. In addition, the formation of dendrites upon repeated stripping and plating of Zn or Li metal on the anode imposes another challenge for their cycling stability. For non-aqueous RFBs, including the RFLBs, the challenge lies in the lack of proper ion-conducting membranes [53]. With the Li^+ ion as an example, the conductivity of Li^+ in the reported membranes is generally 2 orders of magnitude lower than those used in aqueous systems, which severely impairs the power capability of the system and eventually raises the cost. Given the poor power density of non-aqueous flow battery, despite that great efforts have been taken on the development of superior ion-conducting membranes, the near-term deployment of RFBs has to rely on the aqueous systems.

The redox targeting concept represents an implementable solution to significantly increase the energy density of RFBs (Figure 9). Among the recently reported RFB systems, RFLB presents a theoretical energy density as high as 500 Wh/L. However, considering the poor power performance, greater effort should be put on the aqueous RFLB systems for future study. By applying redox targeting concept to water-based battery chemistry, viable RFLB systems with high energy and power densities are anticipated. In addition, robust redox mediators with matched potential to the solid active materials are desired to reduce the free energy loss of redox targeting reactions. Besides, Huang *et al.* used a single redox mediator to realize both the lithiation and delithiation of $LiFePO_4$, which is a useful model to reduce the complexity of chemistry of RFLB [51].

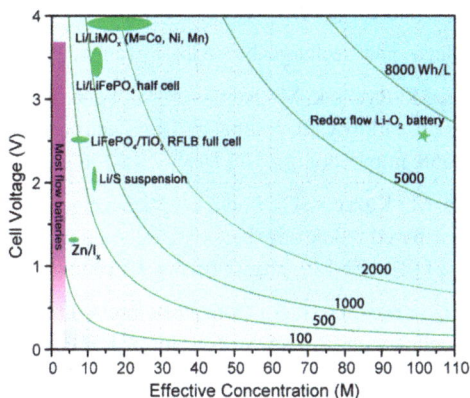

Figure 9. Summary of reported cell voltage, effective concentration and energy density of various aqueous and non-aqueous flow batteries, semi-solid flow batteries and redox flow lithium batteries.

Besides lithium battery, redox targeting concept could be feasibly applied to other battery chemistries. For instance, redox flow sodium-ion battery, which utilizes the earth abundant Na^+ ion storage materials [65] and employs Na^+ as the charge balancing ion, could be developed as a low-cost alternative to RFLB. In addition, redox-assisted ORR and OER reactions deserve systematic investigation. With robust redox catalysts, RFLOB presents an intriguing direction to achieve the stunningly high theoretical energy density (Figure 9).

References

[1]   Soloveichik, G. L., Battery technologies for large-scale stationary energy storage. Annual review of chemical and biomolecular engineering 2011, 2, 503-527. https://doi.org/10.1146/annurev-chembioeng-061010-114116

[2]   Skyllas-Kazacos, M.; Chakrabarti, M. H.; Hajimolana, S. A.; Mjalli, F. S.; Saleem, M., Progress in Flow Battery Research and Development. Journal of The Electrochemical Society 2011, 158 (8), 55. https://doi.org/10.1149/1.3599565

[3]   Díaz-González, F.; Sumper, A.; Gomis-Bellmunt, O.; Villafáfila-Robles, R., A review of energy storage technologies for wind power applications. Renewable

and Sustainable Energy Reviews 2012, 16 (4), 2154-2171. https://doi.org/10.1016/j.rser.2012.01.029

[4]   Thaller, L. H. Electrically rechargeable redox flow cell. 3996064, 1976.

[5]   Skyllas-Kazacos, M.; Rychcik, M.; Robins, R. G.; Fane, A. G.; Green, M. A., New All-Vanadium Redox Flow Cell. Journal of The Electrochemical Society 1986, 133 (5), 1057-1058. https://doi.org/10.1149/1.2108706

[6]   Skyllas-Kazacos, M.; Kazacos, G.; Poon, G.; Verseema, H., Recent advances with UNSW vanadium-based redox flow batteries. International Journal of Energy Research 2010, 34 (2), 182-189. https://doi.org/10.1002/er.1658

[7]   Ponce de León, C.; Frías-Ferrer, A.; González-García, J.; Szánto, D. A.; Walsh, F. C., Redox flow cells for energy conversion. Journal of Power Sources 2006, 160 (1), 716-732. https://doi.org/10.1016/j.jpowsour.2006.02.095

[8]   Lim, H. S.; Lackner, A. M.; Knechtli, R. C., Zinc-Bromine Secondary Battery. Journal of The Electrochemical Society 1977, 124 (8), 1154-1157. https://doi.org/10.1149/1.2133517

[9]   Rydh, C. J.; Sandén, B. A., Energy analysis of batteries in photovoltaic systems. Part II: Energy return factors and overall battery efficiencies. Energy conversion and management 2005, 46 (11), 1980-2000. https://doi.org/10.1016/j.enconman.2004.10.004

[10]   Skyllas-Kazacos, M.; Grossmith, F., Efficient vanadium redox flow cell. Journal of the Electrochemical Society 1987, 134 (12), 2950-2953. https://doi.org/10.1149/1.2100321

[11]   Skyllas-Kazacos, M.; Rychcik, M.; Robins, R. G.; Fane, A.; Green, M., New all-vanadium redox flow cell. J. Electrochem. Soc.;(United States) 1986, 133.

[12]   Skyllas-Kazacos, M.; Rychick, M.; Robins, R. All-vanadium redox battery. 4786567, 1988.

[13]   Sum, E.; Rychcik, M.; Skyllas-Kazacos, M., Investigation of the V (V)/V (IV) system for use in the positive half-cell of a redox battery. Journal of Power sources 1985, 16 (2), 85-95. https://doi.org/10.1016/0378-7753(85)80082-3

[14]   Sum, E.; Skyllas-Kazacos, M., A study of the V (II)/V (III) redox couple for redox flow cell applications. Journal of Power sources 1985, 15 (2), 179-190. https://doi.org/10.1016/0378-7753(85)80071-9

[15] Kazacos, M.; Cheng, M.; Skyllas-Kazacos, M., Vanadium redox cell electrolyte optimization studies. Journal of Applied electrochemistry 1990, 20 (3), 463-467. https://doi.org/10.1007/BF01076057

[16] Skyllas-Kazacos, M.; Kasherman, D.; Hong, D.; Kazacos, M., Characteristics and performance of 1 kW UNSW vanadium redox battery. Journal of Power Sources 1991, 35 (4), 399-404. https://doi.org/10.1016/0378-7753(91)80058-6

[17] Rahman, F.; Skyllas-Kazacos, M., Solubility of vanadyl sulfate in concentrated sulfuric acid solutions. Journal of Power Sources 1998, 72 (2), 105-110. https://doi.org/10.1016/S0378-7753(97)02692-X

[18] Skyllas-Kazacos, M.; Peng, C.; Cheng, M., Evaluation of precipitation inhibitors for supersaturated vanadyl electrolytes for the vanadium redox battery. Electrochemical and solid-state letters 1999, 2 (3), 121-122. https://doi.org/10.1149/1.1390754

[19] Dunn, B.; Kamath, H.; Tarascon, J.-M., Electrical energy storage for the grid: a battery of choices. Science 2011, 334 (6058), 928-935. https://doi.org/10.1126/science.1212741

[20] Chen, J.-q.; Wang, Q.; Wang, B.-g., Research progress in key materials for all vanadium redox flow battery. Modern Chemical Industry 2006, 26 (9), 21.

[21] Kim, J.-H.; Kim, K. J.; Park, M.-S.; Lee, N. J.; Hwang, U.; Kim, H.; Kim, Y.-J., Development of metal-based electrodes for non-aqueous redox flow batteries. Electrochemistry Communications 2011, 13 (9), 997-1000. https://doi.org/10.1016/j.elecom.2011.06.022

[22] Hosseiny, S. S.; Saakes, M.; Wessling, M., A polyelectrolyte membrane-based vanadium/air redox flow battery. Electrochemistry Communications 2011, 13 (8), 751-754. https://doi.org/10.1016/j.elecom.2010.11.025

[23] Menictas, C.; Skyllas-Kazacos, M., Performance of vanadium-oxygen redox fuel cell. Journal of Applied Electrochemistry 2011, 41 (10), 1223-1232. https://doi.org/10.1007/s10800-011-0342-8

[24] Lim, H.; Lackner, A.; Knechtli, R., Zinc-bromine secondary battery. Journal of the Electrochemical Society 1977, 124 (8), 1154-1157. https://doi.org/10.1149/1.2133517

[25] Zhang, L.; Lai, Q.; Zhang, J.; Zhang, H., A high-energy-density redox flow battery based on zinc/polyhalide chemistry. ChemSusChem 2012, 5 (5), 867-9. https://doi.org/10.1002/cssc.201100530

[26] Leung, P. K.; Ponce-de-León, C.; Low, C. T. J.; Shah, A. A.; Walsh, F. C., Characterization of a zinc–cerium flow battery. Journal of Power Sources 2011, 196 (11), 5174-5185. https://doi.org/10.1016/j.jpowsour.2011.01.095

[27] Gong, K.; Ma, X.; Conforti, K. M.; Kuttler, K. J.; Grunewald, J. B.; Yeager, K. L.; Bazant, M. Z.; Gu, S.; Yan, Y., A zinc–iron redox-flow battery under $100 per kW h of system capital cost. Energy & Environmental Science 2015, 8 (10), 2941-2945. https://doi.org/10.1039/C5EE02315G

[28] Li, B.; Nie, Z.; Vijayakumar, M.; Li, G.; Liu, J.; Sprenkle, V.; Wang, W., Ambipolar zinc-polyiodide electrolyte for a high-energy density aqueous redox flow battery. Nature Communication 2015, 6, 6303. https://doi.org/10.1038/ncomms7303

[29] Huskinson, B.; Marshak, M. P.; Suh, C.; Er, S.; Gerhardt, M. R.; Galvin, C. J.; Chen, X.; Aspuru-Guzik, A.; Gordon, R. G.; Aziz, M. J., A metal-free organic-inorganic aqueous flow battery. Nature 2014, 505 (7482), 195-198. https://doi.org/10.1038/nature12909

[30] Lin, K.; Chen, Q.; Gerhardt, M. R.; Tong, L.; Kim, S. B.; Eisenach, L.; Valle, A. W.; Hardee, D.; Gordon, R. G.; Aziz, M. J.; Marshak, M. P., Alkaline quinone flow battery. Science 2015, 349 (6255), 1529-1532. https://doi.org/10.1126/science.aab3033

[31] Lin, K.; Gómez-Bombarelli, R.; Beh, E. S.; Tong, L.; Chen, Q.; Valle, A.; Aspuru-Guzik, A.; Aziz, M. J.; Gordon, R. G., A redox-flow battery with an alloxazine-based organic electrolyte. Nature Energy 2016, 1, 16102. https://doi.org/10.1038/nenergy.2016.102

[32] Herr, T.; Noack, J.; Fischer, P.; Tübke, J., 1,3-Dioxolane, tetrahydrofuran, acetylacetone and dimethyl sulfoxide as solvents for non-aqueous vanadium acetylacetonate redox-flow-batteries. Electrochimica Acta 2013, 113, 127-133. https://doi.org/10.1016/j.electacta.2013.09.055

[33] Shinkle, A. A.; Pomaville, T. J.; Sleightholme, A. E. S.; Thompson, L. T.; Monroe, C. W., Solvents and supporting electrolytes for vanadium acetylacetonate flow batteries. Journal of Power Sources 2014, 248, 1299-1305. https://doi.org/10.1016/j.jpowsour.2013.10.034

[34] Ding, F.; Xu, W.; Graff, G. L.; Zhang, J.; Sushko, M. L.; Chen, X.; Shao, Y.; Engelhard, M. H.; Nie, Z.; Xiao, J.; Liu, X.; Sushko, P. V.; Liu, J.; Zhang, J. G., Dendrite-free lithium deposition via self-healing electrostatic shield mechanism.

Journal of the American Chemical Society 2013, 135 (11), 4450-6.
https://doi.org/10.1021/ja312241y

[35] Wang, Y.; He, P.; Zhou, H., Li-Redox Flow Batteries Based on Hybrid
      Electrolytes: At the Cross Road between Li-ion and Redox Flow Batteries.
      Advanced Energy Materials 2012, 2 (7), 770-779.
      https://doi.org/10.1002/aenm.201200100

[36] Zhao, Y.; Ding, Y.; Song, J.; Peng, L.; Goodenough, J. B.; Yu, G., A reversible Br
      2/Br− redox couple in the aqueous phase as a high-performance catholyte for
      alkali-ion batteries. Energy & Environmental Science 2014.
      https://doi.org/10.1039/c4ee00407h

[37] Zhao, Y.; Wang, L.; Byon, H. R., High-performance rechargeable lithium-iodine
      batteries using triiodide/iodide redox couples in an aqueous cathode. Nature
      communications 2013, 4, 1896. https://doi.org/10.1038/ncomms2907

[38] Zhao, Y.; Byon, H. R., High-Performance Lithium-Iodine Flow Battery. Advanced
      Energy Materials 2013, 3 (12), 1630-1635.
      https://doi.org/10.1002/aenm.201300627

[39] Zhao, Y.; Hong, M.; Bonnet Mercier, N.; Yu, G.; Choi, H. C.; Byon, H. R., A 3.5
      V lithium-iodine hybrid redox battery with vertically aligned carbon nanotube
      current collector. Nano letters 2014, 14 (2), 1085-92.
      https://doi.org/10.1021/nl404784d

[40] Zhao, Y.; Ding, Y.; Song, J.; Li, G.; Dong, G.; Goodenough, J. B.; Yu, G.,
      Sustainable Electrical Energy Storage through the Ferrocene/Ferrocenium Redox
      Reaction in Aprotic Electrolyte. Angewandte Chemie International Edition 2014,
      n/a-n/a.

[41] Yang, Y.; Zheng, G.; Cui, Y., A membrane-free lithium/polysulfide semi-liquid
      battery for large-scale energy storage. Energy & Environmental Science 2013, 6
      (5), 1552-1558. https://doi.org/10.1039/c3ee00072a

[42] Wei, X.; Xu, W.; Vijayakumar, M.; Cosimbescu, L.; Liu, T.; Sprenkle, V.; Wang,
      W., TEMPO-Based Catholyte for High-Energy Density Nonaqueous Redox Flow
      Batteries. Adv Mater 2014, 26 (45), 7649-53.
      https://doi.org/10.1002/adma.201403746

[43] Duduta, M.; Ho, B.; Wood, V. C.; Limthongkul, P.; Brunini, V. E.; Carter, W. C.;
      Chiang, Y.-M., Semi-Solid Lithium Rechargeable Flow Battery. Advanced Energy
      Materials 2011, 1 (4), 511-516. https://doi.org/10.1002/aenm.201100152

[44] Chen, H.; Zou, Q.; Liang, Z.; Liu, H.; Li, Q.; Lu, Y.-C., Sulphur-impregnated flow cathode to enable high-energy-density lithium flow batteries. Nat Commun 2015, 6. https://doi.org/10.1038/ncomms6877

[45] Chen, H.; Lu, Y. C., A High-Energy-Density Multiple Redox Semi-Solid-Liquid Flow Battery. Advanced Energy Materials 2016. https://doi.org/10.1002/aenm.201502183

[46] Fan, F. Y.; Woodford, W. H.; Li, Z.; Baram, N.; Smith, K. C.; Helal, A.; McKinley, G. H.; Carter, W. C.; Chiang, Y. M., Polysulfide flow batteries enabled by percolating nanoscale conductor networks. Nano letters 2014, 14 (4), 2210-8. https://doi.org/10.1021/nl500740t

[47] Wang, Q.; Zakeeruddin, S. M.; Wang, D.; Exnar, I.; Gratzel, M., Redox targeting of insulating electrode materials: a new approach to high-energy-density batteries. Angewandte Chemie 2006, 45 (48), 8197-200. https://doi.org/10.1002/anie.200602891

[48] Huang, Q.; Li, H.; Gratzel, M.; Wang, Q., Reversible chemical delithiation/lithiation of LiFePO4: towards a redox flow lithium-ion battery. Physical chemistry chemical physics : PCCP 2013, 15 (6), 1793-7. https://doi.org/10.1039/C2CP44466F

[49] Pan, F.; Yang, J.; Huang, Q.; Wang, X.; Huang, H.; Wang, Q., Redox Targeting of Anatase TiO2 for Redox Flow Lithium-Ion Batteries. Advanced Energy Materials 2014, 4 (15). https://doi.org/10.1002/aenm.201400567

[50] Pan, F.; Huang, Q.; Huang, H.; Wang, Q., High-Energy Density Redox Flow Lithium Battery with Unprecedented Voltage Efficiency. Chemistry of Materials 2016, 28 (7), 2052-2057. https://doi.org/10.1021/acs.chemmater.5b04558

[51] Huang, Q.; Yang, J.; Ng, C. B.; Jia, C.; Wang, Q., Redox Flow Lithium Battery Based on the Redox Targeting Reactions between LiFePO4 and Iodide. Energy & Environmental Science 2016, 9 (3), 917-921. https://doi.org/10.1039/C5EE03764F

[52] Li, J.; Yang, L.; Yang, S.; Lee, J. Y., The Application of Redox Targeting Principles to the Design of Rechargeable Li–S Flow Batteries. Advanced Energy Materials 2015, 5 (24), n/a-n/a.

[53] Jia, C.; Pan, F.; Zhu, Y. G.; Huang, Q.; Lu, L.; Wang, Q., High–energy density nonaqueous all redox flow lithium battery enabled with a polymeric membrane. Science Advances 2015, 1 (10), e1500886. https://doi.org/10.1126/sciadv.1500886

[54] Chase, G. V.; Zecevic, S.; Wesley, W. T.; Uddin, J.; Sasaki, K. A.; Vincent, G. P.; Bryantsev, V.; Blanco, M.; Addison, D. D., SOLUBLE OXYGEN EVOLVING CATALYSTS FOR RECHARGEABLE METAL-AIR BATTERIES. US Patent 20,120,028,137: 2012.

[55] Chen, Y.; Freunberger, S. A.; Peng, Z.; Fontaine, O.; Bruce, P. G., Charging a Li–O2 battery using a redox mediator. Nature chemistry 2013, 5 (6), 489-494. https://doi.org/10.1038/nchem.1646

[56] Lacey, M. J.; Frith, J. T.; Owen, J. R., A redox shuttle to facilitate oxygen reduction in the lithium air battery. Electrochemistry Communications 2013, 26, 74-76. https://doi.org/10.1016/j.elecom.2012.10.009

[57] Lim, H. D.; Song, H.; Kim, J.; Gwon, H.; Bae, Y.; Park, K. Y.; Hong, J.; Kim, H.; Kim, T.; Kim, Y. H., Superior Rechargeability and Efficiency of Lithium–Oxygen Batteries: Hierarchical Air Electrode Architecture Combined with a Soluble Catalyst. Angewandte Chemie International Edition 2014, 53 (15), 3926-31. https://doi.org/10.1002/anie.201400711

[58] Gao, X.; Chen, Y.; Johnson, L.; Bruce, P. G., Promoting solution phase discharge in Li-O2 batteries containing weakly solvating electrolyte solutions. Nat Mater 2016, 15 (8), 882-888. https://doi.org/10.1038/nmat4629

[59] Sun, D.; Shen, Y.; Zhang, W.; Yu, L.; Yi, Z.; Yin, W.; Wang, D.; Huang, Y.; Wang, J.; Wang, D., A Solution-Phase Bifunctional Catalyst for Lithium-oxygen Batteries. Journal of the American Chemical Society 2014, 136 (25), 8941-6. https://doi.org/10.1021/ja501877e

[60] Imanishi, N.; Luntz, A. C.; Bruce, P., The lithium air battery: fundamentals. Springer: 2014. https://doi.org/10.1007/978-1-4899-8062-5

[61] He, P.; Wang, Y.; Zhou, H., A Li-air fuel cell with recycle aqueous electrolyte for improved stability. Electrochemistry Communications 2010, 12 (12), 1686-1689. https://doi.org/10.1016/j.elecom.2010.09.025

[62] Wang, Y.; He, P.; Zhou, H., Li-Redox Flow Batteries Based on Hybrid Electrolytes: At the Cross Road between Li-ion and Redox Flow Batteries. Advanced Energy Materials 2012, 2 (7), 770-779. https://doi.org/10.1002/aenm.201200100

[63] Zhu, Y. G.; Jia, C.; Yang, J.; Pan, F.; Huang, Q.; Wang, Q., Dual redox catalysts for oxygen reduction and evolution reactions: towards a redox flow Li-O2 battery.

Chem Commun (Camb) 2015, 51 (46), 9451-4.
https://doi.org/10.1039/C5CC01616A

[64] Zhu, Y. G.; Wang, X.; Jia, C.; Yang, J.; Wang, Q., Redox-Mediated ORR and OER Reactions: Redox Flow Lithium Oxygen Batteries Enabled with a Pair of Soluble Redox Catalysts. ACS Catalysis 2016, 6191-6197.
https://doi.org/10.1021/acscatal.6b01478

[65] Kim, S.-W.; Seo, D.-H.; Ma, X.; Ceder, G.; Kang, K., Electrode Materials for Rechargeable Sodium-Ion Batteries: Potential Alternatives to Current Lithium-Ion Batteries. Advanced Energy Materials 2012, 2 (7), 710-721.
https://doi.org/10.1002/aenm.201200026

CHAPTER 9

Advanced Supercapacitors Using Carbon Nanotubes

Hai M. Duong[1], Hanlin Cheng[1], Daniel Jewell[2]

[1]National University of Singapore, Department of Mechanical Engineering, Singapore

mpedhm@nus.edu.sg

[2]University of Cambridge, Department of Materials Science & Metallurgy, UK

daniel.jewell@cantab.net

Abstract

This chapter provides a review of supercapacitors that employ carbon nanotubes (CNT) as electrode materials. We begin with a brief introduction to supercapacitors including their working principles, functional components and key characteristics. We then examine the application of carbon nanotubes in the field of supercapacitors, including discussion on CNTs with zero (0D), one (1D), two (2D) and three-dimensional (3D) structures, and their influence on the electrochemical performance of supercapacitors. Finally, we provide an outlook for the future development and use of carbon nanotubes in supercapacitor applications.

Keywords

Electrochemical Energy Storage, Supercapacitors, Carbon Nanotubes, Carbon Nanotube Composites, Multi-dimensional Structures

Contents

1.   Introduction to Supercapacitors

Electrochemical energy storage (EES) devices are convenient energy units for portable electronic devices, energy backup systems and electrical vehicles. Electrochemical capacitors, commonly referred to as supercapacitors, are a class of EES devices possessing higher energy density than dielectric capacitors and higher power density than traditional batteries. They play a crucial role in building hierarchical energy systems by bridging the gap between dielectric capacitors and chemical batteries.

1.1   Working Principles of Supercapacitors

There are two basic working principles and hence energy storage methods for supercapacitors. The first is the electrochemical double layer capacitor (EDLC) which utilizes physical adsorption of ions at the electrode/electrolyte interface. The working mechanism of an EDLC is quite straightforward. During the operation of an EDLC, a negatively charged anode attracts the cations in the electrolyte to the electrode surface, where they form an electrochemical double layer. Similar physical reactions also occur at the cathode with oppositely charged ions (anions) to maintain the electrostatic balance of the device. The energy stored in the electrode is dependent on the potential difference and charge density in this electrochemical double layer, as illustrated in Figure 1. Due to the similarity of this mechanism with dielectric capacitors, the capacitance of the double layer can be calculated with the following equation:

$$C = \frac{\varepsilon_0 \varepsilon_r A}{d}$$

(1)

where C (F) is the capacitance, ε_0 (F m^{-1}) is the vacuum permitivity, ε_r is the relative permitivity, A is the surface area (m^2) and d (m) is the thickness of the electrochemical double layer.

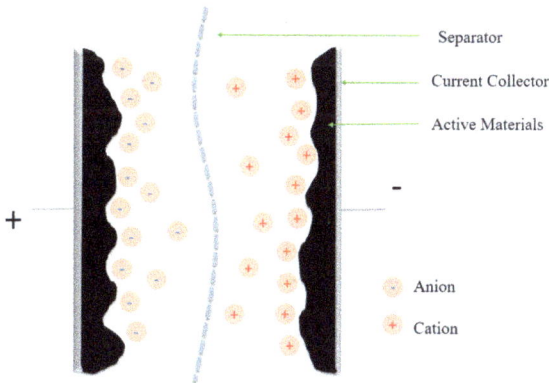

Figure 1 Illustration of charge storage in an electrochemical double layer.

The capacitance of supercapacitors is much larger than that of comparable dielectric capacitors due to the significantly smaller distance, d, which is in the range of several angstroms. It is also important to note that since only physical reactions occur during device operation, there is no chemical or structural change of the electrodes and this endows supercapacitors with near infinite cycling lives.

The second type of supercapacitor is the pseudocapacitor. In pseudocapacitors, faradaic electrochemical reactions are involved to store charges. Despite their thermodynamic similarities to batteries, they undergo surface or near-surface rapid redox reactions and give similar charge/discharge profiles to EDLCs. Because both surface and near-surface materials contribute to the total capacitance, they present higher energy density compared with that of the EDLCs. However, the structural variation and associated volume change of pseudocapacitor electrodes during charge/discharge often lead to a limited cycling life, typically fewer than 10,000 charge/discharge cycles.

In order to evaluate the electrochemical performance of a supercapacitor a number of key parameters need to be assessed, including specific capacitance, operation potential, cycling stability, energy performance, power performance, leakage current and open circuit potential drop: of these parameters, specific capacitance is the main determiner of device performance. Although the majority of reports provide specific capacitance values in gravimetric terms (i.e. based on the mass of electrode materials), it is also important for practical applications to take into account the volumetric and areal capacitance (i.e. based on electrode volume and apparent surface area). To calculate energy performance, Eq. (2) can be used for double layer capacitors with linear charge/discharge profiles, where C is the capacitance, V is the operation potential and E is the stored energy. However, for materials that present strong redox behaviors, an integration should be obtained between accumulated charges and potentials for the energy calculation due to their non-linear charge/discharge profile.

$$E = \frac{1}{2}CV^2 \qquad\qquad (2)$$

1.2   Components of Supercapacitors

The components of a supercapacitor device are similar to those in batteries and include current collectors, electrodes, electrolytes and separators. Each component serves a purpose for which particular properties are essential for proper and efficient operation.

Current collectors: These function to transport electrons from an outer circuit to the electrodes. Although their primary requirement is that of high electrical conductivity, other factors, such as electrode weight, corrosion resistance and mechanical rigidity, should be taken into consideration when selecting current collectors.

Electrodes: These are the capacitance contributors to the device and their surface area and electrochemical activity are the primary selection factors. Both symmetric (two identical electrodes) and asymmetric (two different electrodes) configurations can be employed to compose a supercapacitor full cell.

Electrolytes: These are the ion carriers in the cell, and ionic conductivity and chemical stability are the key parameters that determine device performance. Aqueous electrolytes typically have high ionic conductivities but low operation potential; less than 1.23 V before decomposition of the electrolyte begins. Conversely, organic electrolytes usually have a higher working potential range (commonly >2 V) which produces a larger energy density. However, the ionic conductivity of organic electrolytes is normally one to several orders of magnitude smaller than that of the aqueous electrolytes, resulting in poorer power performance. Further, the flammable nature of organic electrolytes presents additional safety concerns.

Separator: These are barriers between cathode and anode to avoid the circuit shortcut. Mesh size, chemical stability and electrolyte wettability should be taken into consideration when selecting certain types of separator for efficient ion diffusion. For supercapacitors using neutral aqueous electrolyte, cellulose porous membrane can be employed as separator due to its good wettability. When using alkaline aqueous electrolyte, polypropylene or polyvinylidene fluoride are primary selection for their chemical stability. Supercapacitors using organic electrolyte typically employ glass fiber membranes as separators.

1.3 Active Materials for Supercapacitors

Examination of Eq. 1 shows that surface area is a critical factor determining the performance of an EDLC. With this in mind, activated carbon is widely used for EDLC electrode materials due to their high specific surface area ($1,000 \sim 2,000$ m^2 g^{-1}). A wide range of carbon sources can be activated by etching a defected or amorphous carbon material using carbon dioxide, air or the alkaline chemicals, resulting in a structure with high porosity. However, it should be noted that high surface area is not the only requirement for the improvement of EDLC performance. Chmiola *et al.* [2] found that there is an upper limit for the surface area based improvements beyond which further increases cannot improve the specific capacitance, due to the mismatch between the pore size of the carbon electrode and the ion size in the electrolyte. However, recent research has also discovered an anomalous capacitance increase with a decrease of pore size beyond that upper limit [3]. When the capacitance increase diminishes with the pore size decrease, a raised operation potential has brought a sudden capacitance increase. It is proposed that the loss of the solvent shell from the ions may contribute to this extra capacitance increase. Wu *et al.* [4] discovered a periodic changes of the capacitance when decreasing the pore size, which can be attributed to the electrical imaging force from the charges accumulated at electrode/electrolyte interface.

Proper hierarchical pore structures are important to EDLCs. Among the various different methods [5], template synthesis is a direct way to make hierarchical carbon structures. Lithography and other patterning techniques can also be used for fabricating optimized carbon electrodes [6]. Aside from the development of better structures of carbon materials, it is also highly desirable to convert bio-waste to electrode materials to establish an environmentally friendly energy cycle. Currently, nutshell [7], banana peel [8] and crab shell [9] have been successfully made into carbon materials for EDLC applications. In addition to activated carbon, other carbon based materials with high surface area, such as graphene and carbon nanotubes (CNTs), also have been intensively investigated in the field of supercapacitors.

Carbon based materials are widely used for EDLC electrodes, demonstrating a long cycling stability and remarkable power performance. Unfortunately, they suffer a low specific capacitance with limited energy density. Introducing redox reactions to supercapacitor electrodes brings extra pseudocapacitance, which is significant in improving the energy performance of supercapacitors. For this reason, electrodes of metal oxides, metal hydroxides, metal sulfides and conducting polymers have attracted great interest due to their pseudocapactive properties. Unfortunately, metal complex materials typically possess low electrical conductivity, and the structural changes that occur during charge/discharge cycles compromise both the power rate and cycling stability of the device. For conducting polymers, the swelling of molecular chains due to ion insertion upon operation also negatively affects the cycling stability of the device. Ultimately there is a trade-off between power performance, energy performance and cycling stability, and the selection of electrode materials should be carefully considered and tailored to cater to the target application.

To balance the energy and power performance of electrode materials, considerable efforts have been made to develop composite electrodes to leverage the merits of pseudocapacitive and EDLC materials simultaneously. The central concept is to use a carbon host to provide conducting pathways and to act as a structural buffer for the pseudocapacitive guest materials which contribute significant capacitance. From this perspective, various carbon allotropes such as onion carbon [10], CNTs [11] and graphene [12] can be used for composite designs with satisfactory electrochemical performance. Another feasible method to improve energy density of supercapacitors is to dope the carbon with hetero-atoms such as boron and nitrogen. It should be noted that the content ratio between EDLC and psudocapacitive materials must be carefully tuned to meet both a power and energy requirement for a given device: otherwise, the result is simply a mediocre battery [13].

2.   CNTs for Supercapacitors

CNTs are an allotrope of carbon endowed with remarkable mechanical strength, electrical conductivity and chemical stability. They can be regarded as rolled graphene cylinders and are categorized into Single-Walled Nano-Tubes (SWNTs) and Multi-Walled Nano-Tubes (MWNTs) depending on the number of their walls. SWNTs can be further classified to metallic and semiconducting nanotubes based on the wrapping direction of graphitic walls (chirality). MWNTs present metallic elerical properties due the inter-wall interaction. CNTs are generally synthesized using a bottom-up approach such as arc-discharge, laser ablation, or chemical vapor deposition (CVD). Of these techniques, CVD has attracted the most attention due to its reproducibility and high production yield. By considering the structure of CNTs inside the electrode, they can be classified as zero-dimensional (0D), one-dimensional (1D), two-dimensional (2D) or three-dimensional (3D).

Table 1. Classification of different CNT structures for electrodes.

CNT structures	Representatives	Features
0D	CNT particle	Aggregated CNT, binder usage for electrode
1D	CNT fiber	Continuous CNT strand
2D	CNT film	Interconnected horizontal CNT network
	CNT forest	Aligned vertical CNTs
3D	CNT Sponge, foam, aerogel	CNT monolith with interconnected CNTs along 3D space

2.1 0D CNT Sturcture for Supercapacitors

In 0D structures, the surface area of the CNTs is strongly dependent on the number of the walls. For a SWNT, the theoretical surface area is half that of the graphene, given that the inner surface is inaccessible. As the number of walls increases, the MWNT specific surface area dramatically decreases. Consequently, MWNTs possessing more than 10 walls usually have specific capacitance less than 40 F g^{-1}. While for SWNTs, specific capacitance more than 80 F g^{-1} can be reached. Besides the specific capacitance, other excellent properties including robust mechanical rigidity and high electrical conductivity render them potential candidates for supercapacitor applications.

0D CNT usually refers to use of entangled CNTs in the form of powders. To make them into electrochemical electrodes, additives such as polymer binders and ethylene carbon have to be applied. Typical supercapacitor electrodes using pristine CNTs have specific capacitance no more than 50 F g^{-1} [14,15]. Various methods, including chemical functionalization [16], chemical activation [17] and hetero-atom doping [18], have been explored to improve the capacitive performance of CNTs by introducing redox reaction centers which extra pseudocapacitive capacitance. As resules, specific capacitance more than 100 F g^{-1} can be reached. The main application of this type of electrode is for alternating current (AC) linear filtering, rather than energy storage. However, CNTs still struggle to compete with active carbon with regard to energy density due to their relatively small surface area (< 200 m^2 g^{-1}) compared to activated carbons (1,000 ~ 2,000 m^2 g^{-1}).

2.2 1D CNT Structure for Supercapacitors

In order to remove the hazardous additives from the electrodes. CNT based binder-free electrodes have also been developed. 1D CNT fibers are generally fabricated from dry-spinning [19], wet spinning [20] or a continuous CVD process [21]. Using the CVD technique, it is possible to prepare high density CNT forests on Si wafers. For the dry-spinning method, the CNTs can be directly spun to form fibers due to the strong van der Waals interaction between nanotubes. However, in order to achieve a continuous fabrication of CNT fibers with a certain length, the spinning speed is very limited, which leads to a low production yield. In contrast, the wet-spinning method is carried out to spin CNT/polymer composites into fiber form with a high production yield. The polymer residues inside can be subsequently removed using calcination or proper etching. Sun *et al.* [22] have used CNT fibers directly as supercapacitor electrodes. When measured with a three-electrode configuration, these CNT fibers present a small specific capacitance of < 5 F g^{-1}. However, with simple electrochemical activation in H$_2$SO$_4$ and Na$_2$SO$_4$ solutions, the capacitance can be promoted to over 100 F g^{-1}.

In addition to improving the electrodes, introducing redox species into the electrolyte is also an important strategy to increase the electrochemical performance of supercapacitors. Xu *et al.* [23] have attempted the introduction of hydroquinone into a H$_2$SO$_4$/PVA electrolyte. The redox reactions coming from the hydroquinone promote the volumetric capacitance and result in an increase from 17 to 42 F cm^{-3}. This device produced using redox mediates also exhibits a good cycling stability of over 2,000 cycles. For fiber based CNT structures, the loading of the active materials can only be realized along one dimension. For a practical device requiring high energy this would require a large number of these fiber-based composites, leading to a more complex system structure.

2.3 2D CNT Structure for Supercapacitors

In addition to the 1D CNT structures, 2D film-based CNT structures also have been developed for supercapacitor applications. A convenient method to obtain CNT films is to filtrate the CNT dispersion through a porous membrane to form horizontally located

CNTs. In this method, the homogeneity of the dispersion plays a key role in determining the electrode performance because distribution of CNTs with uniform interconnectivity is important for both electron transport and ion diffusion. It is found that CNT films synthesized from this method can deliver a capacitance of 55 F g^{-1} [24]. Besides the vacuum filtration, spray coating CNT dispersion onto metallic substrate has also been used to fabricate CNT thin films [25] with capacitance values of over 100 F g^{-1}. The direct electrical contact between the conducting contact and CNTs films are believed to improve the specific capacitances. But the usage of heavy metallic current collector inevitably increase the weight of the supercapacitor device.

For CNT films synthesized from a solution route, major challenges exist in achieving homogeneity with minimal structural damage to the CNTs. Considering the hydrophobicity of the CNTs, it is difficult to disperse them in proper solvent in order to obtain films with optimized CNT interconnection. Consequently, there has been a growing interest in the preparation of such films via drying processes to avoid the solution stage during synthesis. CNT films can also be directly obtained from the floating catalytic chemical deposition (FCCVD) technique. In a typical FCCVD process [26], carbon feedstock together with a catalyst (ferrocene) is injected into a furnace. Under the reaction temperature, the ferrocene is decomposed to give iron particles, which serve as nucleation centers for CNT growth. Due to the high density of CNTs inside the furnace, the CNTs interconnect and attach to the substrate or chamber walls in the form of films. Supercapacitor electrodes have been manufactured using these films and it has also been reported that increasing either working temperature or compressive stress on such films can improve their specific capacitance (i. e. from 25 to 50 F g^{-1} under a compressive pressure of 1,000 kPa) [27].

In addition to film structures with horizontally oriented CNTs, vertically aligned CNTs have also been synthesized and utilized for supercapacitor applications. Due to the high density of the CNTs, they are sometimes referred as CNT "forests" [28]. These CNT "forests" are synthesized via a CVD process and their quality is highly dependent on the distribution of catalyst particles on the substrate. Unfortunately, traditional CVD processes usually lead to inter-tube entanglement and small nanotube length, which results in poor mechanical strength and low electrical conductivity. More recently, water-assisted CVD and plasma enhanced chemical vapor deposition (PECVD) have been developed to improve the CNT properties to meet the demands of electrochemical applications. It is found that CNTs synthesized from PECVD can deliver a specific capacitance of over 15 F g^{-1} [29]. To further improve the capacitance value, a multi-catalyst system has been applied in CNT forest synthesis, delivering a specific capacitance of over 60 F g^{-1} [30].

2.4   3D CNT Structure for Supercapacitors

3D CNT structures attract great attention for supercapacitor application due to their 3D electron transport and diffusion channels. Initially, the main focus of research was on the synthesis of 3D CNT structures using 3D scaffolds. For example, Fang et al. [31] firstly fabricated an electrically conducting scaffold using nickel fiber and then coated this nickel scaffold with a polymer/CNT dispersion. The thermally annealed 3D carbon/CNT composites have achieved a specific capacitance of over 200 F g^{-1}. However, the dispersion coating method leads to a poor interface between the CNTs and 3D conducting substrate, leading to small charge current no more than 0.5 A g^{-1}.

As practical devices, the total weight of supercapacitors have been taken into consideration when estimating its volumetric and gravimetric energy density including the current collectors, active materials, separator and other components. For 3D CNT relying on a metallic scaffold, the device as a whole presents a low gravimetric energy due to the large mass contribution of the scaffold. It is therefore highly desirable to develop a self-supported free-standing 3D CNT structure where the CNT network can serve the role of both current collector and active material. Based on this concept, a CNT sponge was fabricated using the FFCVD system, exhibiting good mechanical flexibility and low electrical resistivity of 6×10^{-3} Ω m [32]. More recently, greenhouse gas decomposition methods have allowed carbon dioxide to be converted to a CNT foam for supercapacitor applications [33]. It is worth noting that this FCCVD process is non-continuous and therefore the CNT foam inside the reaction chamber has to be removed in time for the next batch of synthesis

Solution synthesis methods can be employed to fabricate 3D CNT structures. It is found that functionalized CNTs in certain solvents can from 3D monoliths with the help of molecular crosslinkers. Zou et al. [34] decorated CNTs with 3-(trimethoxysilyl) propyl methacrylate which connects CNTs together to form a self-supported structure. Similarly, functional molecules such as ferrocene [35], fibroin [36], and even viruses [37] can be used as crosslinkers to obtain 3D structures of CNTs. Compared with the 3D CNTs obtained from the FCCVD method, these solution-derived 3D CNTs structures are composed of mesopores with an average pore size from 2 to 50 nm. Therefore, in order to avoid structural damage from the normal drying process, they are either freeze- or supercritically-dried.

3. CNT Composites for Supercapacitors

3.1 0D CNT Composites for Supercapacitors

Incorporating CNTs with pseudocapacitive materials, including metal oxides, metal sulfides, metal hydroxides, and conducting polymers, is an effective method to raise device performance. For example, $Ni(OH)_2$/CNT [38], NiO/CNT [39], $Co(OH)_2$/CNT [40], Co_3O_4/CNT [41] can be synthesized via a precipitation method to deposit metal oxides/hydroxides on CNTs with specific capacitances from 400 to more than 1,000 F g^{-1}. For MnO_x related CNT composites, both chemical precipitation and in-situ reduction can be utilized to fabricate the composite, taking into account the strong oxidative capability of manganese related chemicals such as $KMnO_4$ [42,43]. The specific capacitance of these composites varies from 100 to more than 1,200 F g^{-1} [44], depending on the utilization of the metal oxides/hydroxides. It is reported that a small mass loading of active materials below 1 mg cm^{-2} in electrodes can drastically increase specific capacitance. This strategy, however, greatly lowers the areal and volumetric capacitance of the electrodes, limiting their wider application. Conducting polymers can, however, easily coat the CNTs based on a facile chemical polymerization process in which the CNTs mediate the swelling and possible damage on the polymer chains during the charge/discharge process. These composites present high capacitance competitive to metal oxides/hydroxides but with the benefit of better electrical conductivity. In the case of CNT composites in the form of powders, polymer additives have to be incorporated inside the electrode along with carbon additives such as acetylene black which compensate for the insulating nature of the polymer binder. The use of a polymer binder and carbon additives inevitably lowers the percentage of capacitive active materials in the device.

3.2 1D CNT Composites for Supercapacitors

For CNT fiber-based supercapacitors, the twisted strands cause serious bundling of the CNTs, limiting the area accessible to the electrolyte and thereby resulting in a small specific capacitance. Various pseudocapacitive components can be introduced into the CNT fibers to improve their capacitive performance. After obtaining the CNT fiber by spinning them out from the CNT forest, MnO_2 can be deposited on the surface using electrochemical deposition [45]. The fabrication of a flexible symmetric supercapacitor has been achieved using two MnO_2/CNT composite fibers twisted together to form a symmetric supercapacitor, with LiCl/PVA hydrogel used as both electrolyte and separator. The prepared device has achieved a volumetric capacitance of 34.6 F cm^{-3} whilst maintaining a capacitance strain of 84% at a tension strain of 37.5%. More significantly, Wang et al. [46] have developed a one-step synthesis and device fabrication process via electrochemical deposition. This method is applicable for coating various materials, such as MnO_2, polyaniline and even graphene, on CNT fibers. The prepared devices can reach a volumetric capacitance as high as 68.4 F cm^{-3}.

3.3   2D CNT Composites for Supercapacitors

Interconnected 2D CNT films provide an excellent substrate for deposition of various other active capacitance components. Materials such as Co_3O_4 [47], MnO_2 [48] and $Ni(OH)_2$ [49] have been incorporated into CNT films with specific capacitance of above 1,000 F g^{-1}. More recently, Cheng et al [50] developed a continuous synthesis of CNT films based on a FFCVD technique where CNTs are drawn as long sheets. The generated CNT films and their related $Ni(OH)_2$/CNT film composites showed good electrochemical performance for supercapacitor applications with specific capacitance of over 1,000 F g^{-1}.

In order to further raise the specific capacitance, pseudocapacitive materials such as MnO_2 was incorporated into the CNT forests to produce materials with specific capacitance of over 300 F g^{-1} [33]. The main problem of the 2D CNT composites for supercapacitor applications is its low mass loading of the guest materials, leading to small areal and volumetric capacitance. The further increase of the pseudocapacitive materials brings unwanted aggregations, leading to poor electrochemical performance.

3.4   3D CNT Composites for Supercapacitors

3D CNT composites have also been prepared for supercapacitor applications. Chen et al.[51] coated a commercial available cleaning sponge with a CNT dispersion and later electrochemically deposited the MnO_2. The obtained 3D composite achieved a capacitance as high as 1,350 F g^{-1} based on the mass of MnO_2. Using nickel foam as a scaffold, $Ni(OH)_2$/CNT 3D structures have also been developed with capacitance values over 3,300 F g^{-1} and cycling lives over 3,000 cycles [52]. A free-standing 3D CNT sponge with deposited MnO_x has been found to achieve specific capacitance over 462 F g^{-1} [53]. The content of the MnO_x has been carefully investigated in this structure and the authors pointed out a critical percentage of MnO_x, above which the MnO_2 aggregates together, resulting in a capacitance loss. Using conducting polymers as an intermediate layer, Li et al. [54] fabricated a ternary MnO_2/PPy/CNT for supercapacitor application based on a CNT sponge. Due to the existence of PPy as a middle layer, the structure has a high capacitance over 200 F g^{-1} at a scan rate of 200 mV s^{-1}. Further, the volumetric capacitance is found to increase by 77 % at 50% compression, indicating the possibility to fabricate a compressive supercapacitor for wearable applications.

3D free-standing CNT aerogels have also been fabricated utilizing the strong van der Waals interaction between CNTs [55]. More creatively, Cheng et al. [56,57] have developed a CNT hydrogel as well as a xerogel for supercapacitor applications. Perhaps counterintuitively, it has been found that seriously bundled CNTs in xerogels present higher specific capacitance. The better performance can be attributed to the conductivity of larger CNT bundles. The authors also point out that the structural difference between CNT hydrogels and xerogels plays a major role in determining the electrochemical performance of delicately designed $Ni(OH)_2$/CNT and Mn_3O_4/CNT composite gels. Compared with the core-shell structure presented in the CNT hydrogel composite, the nest structure in the CNT xerogel composite exhibits better cycling stability because $Ni(OH)_2$ is wrapped by a CNT nest, accommodating the volume expansion during cycling.

4.   Outlook

The use of pristine CNTs as electrode materials for supercapacitors presents drawbacks of low specific capacitance and consequently low gravimetric energy density. However, the high electrical conductivity makes them good candidates for auxiliary power in energy supply systems. They have also shown promise for linear filtrating units, demonstrating great potential in replacing traditional electrolytic capacitors [58].

Combining CNTs with advances in pseudocapacitive materials has become a major research focus [59]. The development of hybrid supercapacitors using insertion-type cathode is now a popular topic [60]. Currently, various asymmetric devices with battery type electrodes as the positive electrode and carbon material as the negative electrode have been reported [61,62]. This design is aimed at achieving a higher energy density for the whole device. However, the low electrical conductivity and poor ion diffusion in solid lithium ion type electrodes may hinder the overall discharge rate. By incorporating CNTs into the positive electrodes, better performance can be achieved.

Another research direction is to develop CNT composites with high dimensional hierarchical structures. To date, powder based composites have been intensively investigated. Efforts should be devoted to the study of novel 3D CNT composites, especially for free-standing structures, to unravel the role of interconnected CNTs in the electrochemical performance of composites with high dimensional structure.

Given the remarkable electrical and mechanical properties, the potential roles of CNTs as electrode materials should not be underestimated. Using CNTs as current collectors can greatly optimize the overall configuration of supercapacitor devices, and significant interest now exists for the development of wearable energy storage devices [63]. Compared with the Li ion battery, CNTs offer a safer and better cycling reliability, whilst the interconnected-CNT structure helps to improve the resistance of devices to large volume changes.

References

[1] A .G. Pandolfo, A. F. Hollenkamp, Carbon properties and their role in supercapacitors, J. Power Sources. 157 (2006) 11–27. https://doi.org/10.1016/j.jpowsour.2006.02.065

[2] J. Chmiola, G. Yushin, R. Dash, Y. Gogotsi, Effect of pore size and surface area of carbide derived carbons on specific capacitance, J. Power Sources. 158 (2006) 765–772. https://doi.org/10.1016/j.jpowsour.2005.09.008

[3] J. Chmiola, Anomalous Increase in Carbon Capacitance at Pore Sizes Less Than 1 Nanometer, Science. 313 (2006) 1760–1763. https://doi.org/10.1126/science.1132195

[4] P. Wu, J. Huang, V. Meunier, B. G. Sumpter, R. Qiao, Complex capacitance scaling in ionic liquids-filled nanopores, ACS Nano. 5 (2011) 9044–9051. https://doi.org/10.1021/nn203260w

[5] J. Wei, D. Zhou, Z. Sun, Y. Deng, Y. Xia, D. Zhao, A controllable synthesis of rich nitrogen-doped ordered mesoporous carbon for CO_2 capture and supercapacitors, Adv. Funct. Mater. 23 (2013) 2322–2328. https://doi.org/10.1002/adfm.201202764

[6] P. Huang, C. Lethien, S. Pinaud, K. Brousse, R. Laloo, V. Turq, et al., On-chip and freestanding elastic carbon films for micro-supercapacitors, Science. 351 (2016) 691–695. https://doi.org/10.1126/science.aad3345

[7] J. Xu, Q. Gao, Y. Zhang, Y. Tan, W. Tian, L. Zhu, et al., Preparing two-dimensional microporous carbon from Pistachio nutshell with high areal capacitance as supercapacitor materials., Sci. Rep. 4 (2014) 5545. https://doi.org/10.1038/srep05545

[8] Y. Lv, L. Gan, M. Liu, W. Xiong, Z. Xu, D. Zhu, et al., A self-template synthesis of hierarchical porous carbon foams based on banana peel for supercapacitor electrodes, J. Power Sources. 209 (2012) 152–157. https://doi.org/10.1016/j.jpowsour.2012.02.089

[9] H. J. Liu, X. M. Wang, W. J. Cui, Y. Q. Dou, D. Y. Zhao, Y. Y. Xia, Highly ordered mesoporous carbon nanofiber arrays from a crab shell biological template and its application in supercapacitors and fuel cells, J. Mater. Chem. 20 (2010) 4223. https://doi.org/10.1039/b925776d

[10] D. Pech, M. Brunet, H. Durou, P. Huang, V. Mochalin, Y. Gogotsi, et al., Ultrahigh-power micrometre-sized supercapacitors based on onion-like carbon, Nat. Nanotechnol. 5 (2010) 651–654. https://doi.org/10.1038/nnano.2010.162

[11] B. Kay, H. An, W. S. Kim, Y. S. Park, J. Moon, D. J. Bae, et al., Electrochemical Properties of High-Power Supercapacitors Using Single-Walled Carbon Nanotube Electrodes, Adv. Funct. Mater. 11 (2001) 387–392. https://doi.org/10.1002/1616-3028(200110)11:5<387::AID-ADFM387>3.0.CO;2-G

[12] M. D. Stoller, S. Park, Z. Yanwu, J. An, R. S. Ruoff, Graphene-Based ultracapacitors, Nano Lett. 8 (2008) 3498–3502. https://doi.org/10.1021/nl802558y

[13] Y. Gogotsi, P. Simon, True Performance Metrics in Electrochemical Energy Storage, Science. 334 (2011) 917–918. https://doi.org/10.1126/science.1213003

[14] A. González, E. Goikolea, J. Andoni, R. Mysyk, Review on supercapacitors : Technologies and materials, 58 (2016) 1189–1206.

[15] P. W. Ruch, R. Kötz, A. Wokaun, Electrochemical characterization of single-walled carbon nanotubes for electrochemical double layer capacitors using non-aqueous electrolyte, Electrochim. Acta. 54 (2009) 4451–4458. https://doi.org/10.1016/j.electacta.2009.03.022

[16] C. Niu, E. K. Sichel, R. Hoch, D. Moy, H. Tennent, High power electrochemical capacitors based on carbon nanotube electrodes, Appl. Phys. Lett. 70 (1997) 1480. https://doi.org/10.1063/1.118568

[17] C. G. Liu, M. Liu, F. Li, H. M. Cheng, Frequency response characteristic of single-walled carbon nanotubes as supercapacitor electrode material, Appl. Phys. Lett. 92 (2008) 67–70. https://doi.org/10.1063/1.2907501

[18] Z. Jiang, A. Al-Zubaidi, S. Kawasaki, Unusual increase in the electric double-layer capacitance with charge–discharge cycles of nitrogen doped single-walled carbon nanotubes, Mater. Express. 4 (2014) 331–336. https://doi.org/10.1166/mex.2014.1174

[19] M. Zhang, K. R. Atkinson, R. H. Baughman, Multifunctional carbon nanotube yarns by downsizing an ancient technology., Science. 306 (2004) 1358–1361. https://doi.org/10.1126/science.1104276

[20] B. Vigolo, A. Pénicaud, C. Coulon, C. Sauder, R. Pailler, C. Journet, et al., Macroscopic fibers and ribbons of oriented carbon nanotubes., Science. 290 (2000) 1331–1334. https://doi.org/10.1126/science.290.5495.1331

221

[21]   Y. L. Li, I. A. Kinloch, A.H. Windle, Direct Spinning of Carbon Nanotube Fibers from Chemical Vapor Deposition Synthesis, Science. 304 (2004) 276–278. https://doi.org/10.1126/science.1094982

[22]   G. Sun, J. Zhou, F. Yu, Y. Zhang, J. H. L. Pang, L. Zheng, Electrochemical capacitive properties of CNT fibers spun from vertically aligned CNT arrays, J. Solid State Electrochem. 16 (2012) 1775–1780. https://doi.org/10.1007/s10008-011-1606-2

[23]   R. Xu, F. Guo, X. Cui, L. Zhang, K. Wang, J. Wei, High performance carbon nanotube based fiber-shaped supercapacitors using redox additives of polypyrrole and hydroquinone, J. Mater. Chem. A. 3 (2015) 22353–22360. https://doi.org/10.1039/C5TA06165B

[24]   P. J. King, T. M. Higgins, S. De, N. Nicoloso, J.N. Coleman, K. E. T. Al, Percolation Effects in Supercapacitors with Thin, Transparent Carbon Nanotube Electrodes, ACS Nano. 6 (2012) 1732–1741. https://doi.org/10.1021/nn204734t

[25]   M. Kaempgen, C.K. Chan, J. Ma, Y. Cui, G. Gruner, Printable thin film supercapacitors using single-walled carbon nanotubes, Nano Lett. 9 (2009) 1872–1876. https://doi.org/10.1021/nl8038579

[26]   H. Zhu, B. Wei, Direct fabrication of single-walled carbon nanotube macro-films on flexible substrates., Chem. Commun. (2007) 3042–3044. https://doi.org/10.1039/b702523h

[27]   X. Li, J. Rong, B. Wei, Electrochemical Behavior of Single-walled Carbon Nanotube Stress, ACS Nano. 4 (2010) 6039–6049. https://doi.org/10.1021/nn101595y

[28]   Li, Xie, Qian, Chang, Zou, Zhou, et al., Large-Scale Synthesis of Aligned Carbon Nanotubes, Science. 274 (1996) 1701–1703. https://doi.org/10.1126/science.274.5293.1701

[29]   D. N. Futaba, K. Hata, T. Yamada, T. Hiraoka, Y. Hayamizu, Y. Kakudate, et al., Shape-engineerable and highly densely packed single-walled carbon nanotubes and their application as super-capacitor electrodes., Nat. Mater. 5 (2006) 987–94. https://doi.org/10.1038/nmat1782

[30]   S. Dörfler, I. Felhősi, T. Marek, S. Thieme, H. Althues, L. Nyikos, et al., High power supercapacitor electrodes based on vertical aligned carbon nanotubes on aluminum, J. Power Sources 227 (2013) 218–228. https://doi.org/10.1016/j.jpowsour.2012.11.068

[31] Y. Fang, F. Jiang, H. Liu, X. Wu, Y. Lu, Free-standing Ni-microfiber-supported carbon nanotube aerogel hybrid, RSC Adv. 2 (2012) 6562–6569. https://doi.org/10.1039/c2ra20271a

[32] X. Gui, J. Wei, K. Wang, A. Cao, H. Zhu, Y. Jia, et al., Carbon Nanotube Sponges, Adv. Mater. 22 (2010) 617–621. https://doi.org/10.1002/adma.200902986

[33] Y. Wang, H. Liu, X. Sun, I. Zhitomirsky, Manganese dioxide-carbon nanotube nanocomposites for electrodes of electrochemical supercapacitors, Scr. Mater. 61 (2009) 1079–1082. https://doi.org/10.1016/j.scriptamat.2009.08.040

[34] J. Zou, J. Liu, A. S. Karakoti, A. Kumar, D. Joung, Q. Li, et al., Ultralight multiwalled carbon nanotube aerogel, ACS Nano. 4 (2010) 7293–7302. https://doi.org/10.1021/nn102246a

[35] R. R. Kohlmeyer, M. Lor, J. Deng, H. Liu, J. Chen, Preparation of stable carbon nanotube aerogels with high electrical conductivity and porosity, Carbon. 49 (2011) 2352–2361. https://doi.org/10.1016/j.carbon.2011.02.001

[36] S. M. Kwon, H. S. Kim, H. J. Jin, Multiwalled carbon nanotube cryogels with aligned and non-aligned porous structures, Polymer. 50 (2009) 2786–2792. https://doi.org/10.1016/j.polymer.2009.04.056

[37] P. Y. Chen, M.N. Hyder, D. Mackanic, N. M. D. Courchesne, J. Qi, M.T. Klug, et al., Assembly of viral hydrogels for three-dimensional conducting nanocomposites., Adv. Mater. 26 (2014) 5101–5017. https://doi.org/10.1002/adma.201400828

[38] Y. Wang, L. Yu, Y. Xia, Electrochemical Capacitance Performance of Hybrid Supercapacitors Based on $Ni(OH)_2$/Carbon Nanotube Composites and Activated Carbon, J. Electrochem. Soc. 153 (2006) A743. https://doi.org/10.1149/1.2171833

[39] A. D. Su, X. Zhang, A. Rinaldi, S. T. Nguyen, H. Liu, Z. Lei, et al., Hierarchical porous nickel oxide-carbon nanotubes as advanced pseudocapacitor materials for supercapacitors, Chem. Phys. Lett. 561-562 (2013) 68–73. https://doi.org/10.1016/j.cplett.2013.01.023

[40] C. Mondal, D. Ghosh, M. Ganguly, A. Kumar, A. Roy, T. Pal, Synthesis of multiwall carbon nanotube wrapped $Co(OH)_2$ flakes : A high-performance supercapacitor, Appl. Surf. Sci. 359 (2015) 500–507. https://doi.org/10.1016/j.apsusc.2015.10.078

[41] L. Tao, L. Shengjun, Z. Bowen, W. Bei, N. Dayong, C. Zeng, et al., Supercapacitor electrode with a homogeneously Co_3O_4-coated multiwalled carbon nanotube for a high capacitance, Nanoscale Res. Lett. 10 (2015) 208. https://doi.org/10.1186/s11671-015-0915-2

[42] H. Zheng, J. Wang, Y. Jia, C. Ma, In-situ synthetize multi-walled carbon nanotubes@MnO_2 nanoflake core-shell structured materials for supercapacitors, J. Power Sources. 216 (2012) 508–514. https://doi.org/10.1016/j.jpowsour.2012.06.047

[43] J. M. Ko, K. M. Kim, Electrochemical properties of MnO_2/activated carbon nanotube composite as an electrode material for supercapacitor, Mater. Chem. Phys. 114 (2009) 837–841. https://doi.org/10.1016/j.matchemphys.2008.10.047

[44] J. Kim, K. H. Lee, L. J. Overzet, G. S. Lee, Synthesis and Electrochemical Properties of Spin-Capable Carbon Nanotube Sheet/MnO, Nano Lett. 11 (2011) 2611–2617. https://doi.org/10.1021/nl200513a

[45] C. Choi, H. J. Sim, G. M. Spinks, X. Lepró, R. H. Baughman, S. J. Kim, Elastomeric and Dynamic MnO_2/CNT Core-Shell Structure Coiled Yarn Supercapacitor, Adv. Energy Mater. 6 (2016) 1–8. https://doi.org/10.1002/aenm.201502119

[46] B. Wang, X. Fang, H. Sun, S. He, J. Ren, Y. Zhang, et al., Fabricating Continuous Supercapacitor Fibers with High Performances by Integrating All Building Materials and Steps into One Process, Adv. Mater. 27 (2015) 7854–7860. https://doi.org/10.1002/adma.201503441

[47] R. R. Salunkhe, K. Jang, S. W. Lee, S. Yu, H. Ahn, Binary metal hydroxide nanorods and multi-walled carbon nanotube composites for electrochemical energy storage applications, J. Mater. Chem. 22 (2012) 21630–21635. https://doi.org/10.1039/c2jm32638h

[48] T. M. Higgins, D. McAteer, J. C. M. Coelho, B. M. Sanchez, Z. Gholamvand, G. Moriarty, et al., Effect of percolation on the capacitance of supercapacitor electrodes prepared from composites of manganese dioxide nanoplatelets and carbon nanotubes, ACS Nano. 8 (2014) 9567–9579. https://doi.org/10.1021/nn5038543

[49] H. Fang, S. Zhang, T. Jiang, R. Lin, Y. Lin, One-step synthesis of Ni/Ni(OH)$_2$@Multiwalled carbon nanotube coaxial nanocable film for high performance supercapacitors, Electrochim. Acta. 125 (2014) 427–434. https://doi.org/10.1016/j.electacta.2014.01.128

[50] H. Cheng, K. L. P. Koh, P. Liu, T. Q. Thang, H.M. Duong, Continuous self-assembly of carbon nanotube thin films and their composites for supercapacitors, Colloids Surfaces A Physicochem. Eng. Asp. 481 (2015) 626–632. https://doi.org/10.1016/j.colsurfa.2015.06.039

[51] W. Chen, R.B. Rakhi, L. Hu, X. Xie, Y. Cui, H. N. Alshareef, High Performance Nanostructured Supercapacitors on a Sponge High Performance Nanostructured Supercapacitors on a Sponge, Nano Lett. 11 (2011) 5165–5172. https://doi.org/10.1021/nl2023433

[52] Z. Tang, C. Tang, H. Gong, A High Energy Density Asymmetric Supercapacitor from Nano-architectured Ni(OH)$_2$/Carbon Nanotube Electrodes, Adv. Funct. Mater. 22 (2012) 1272–1278. https://doi.org/10.1002/adfm.201102796

[53] D. Zhao, Z. Yang, L. Zhang, X. Feng, Y. Zhang, Electrodeposited Manganese Oxide on Nickel Foam–Supported Carbon Nanotubes for Electrode of Supercapacitors, Electrochem. Solid-State Lett. 14 (2011) A93. https://doi.org/10.1149/1.3562927

[54] P. Li, Y. Yang, E. Shi, Q. Shen, Y. Shang, S. Wu, et al., Core-double-shell, carbon nanotube@polypyrrole@MnO$_2$ sponge as freestanding, compressible supercapacitor electrode, ACS Appl. Mater. Interfaces. 6 (2014) 5228–5234. https://doi.org/10.1021/am500579c

[55] M. B. Bryning, D. E. Milkie, M. F. Islam, L. A. Hough, J. M. Kikkawa, A. G. Yodh, Carbon Nanotube Aerogels, Adv. Mater. 19 (2007) 661–664. https://doi.org/10.1002/adma.200601748

[56] H. Cheng, H. M. Duong, Three dimensional carbon nanotube/nickel hydroxide gels for advanced supercapacitors, RSC Adv. 5 (2015) 30260–30267. https://doi.org/10.1039/C5RA01847A

[57] H. Cheng, H. M. Duong, D. Jewell, Three dimensional manganese oxide on carbon nanotube hydrogels for asymmetric supercapacitors, RSC Adv. 6 (2016) 36954–36960. https://doi.org/10.1039/C6RA02858F

[58] Y. Rangom, X. S. Tang, L. F. Nazar, Carbon Nanotube-Based Supercapacitors with Excellent ac Line Filtering and Rate Capability via Improved Interfacial Impedance, ACS Nano. 9 (2015) 7248–7255. https://doi.org/10.1021/acsnano.5b02075

225

[59]   D. Cericola, R. Kötz, Hybridization of rechargeable batteries and electrochemical capacitors: Principles and limits, Electrochim. Acta. 72 (2012) 1–17. https://doi.org/10.1016/j.electacta.2012.03.151

[60]   M. Salanne, B. Rotenberg, K. Naoi, K. Kaneko, P. L. Taberna, C. P. Grey, et al., Efficient storage mechanisms for building better supercapacitors, Nat. Energy. 1 (2016) 16070. https://doi.org/10.1038/nenergy.2016.70

[61]   A. Vlad, N. Singh, J. Rolland, S. Melinte, P. M. Ajayan, J. F. Gohy, Hybrid supercapacitor-battery materials for fast electrochemical charge storage., Sci. Rep. 4 (2014) 4315. https://doi.org/10.1038/srep04315

[62]   W. Zuo, C. Wang, Y. Li, J. Liu, Directly grown nanostructured electrodes for high volumetric energy density binder-free hybrid supercapacitors: a case study of CNTs//$Li_4Ti_5O_{12}$., Sci. Rep. 5 (2015) 7780. https://doi.org/10.1038/srep07780

[63]   L. Liu, Y. Yu, C. Yan, K. Li, Z. Zheng, Wearable energy-dense and power-dense supercapacitor yarns enabled by scalable graphene-metallic textile composite electrodes., Nat. Commun. 6 (2015) 7260. https://doi.org/10.1038/ncomms8260

Keywords

About the Editors

Dr. Li Lu is Full Professor at the Department of Mechanical Engineering, National University of Singapore. His research interests include nanostructure materials, all-solid-state batteries, energy storage materials and other types of functional materials. He is the Editor-in-Chief of Functional Materials Letters, and Associate Editor of Materials Technology.

Dr. Ning Hu is Full Professor and the Dean and of College of Aerospace Engineering, Chongqing University, China. He has worked on a broad range of research topics including Structural and functional composites, Computational solids mechanics, Structural engineering. Recently, his research has been mainly focused on: fabrication and property evaluation of structural and functional nanocomposites; computational material science, e.g., multi-scale simulations of various phenomena of composites and other materials; structural health monitoring and non-destructive damage evaluation techniques; impact behaviors of FRP composites, etc. He is now an Associate Editor, Leading Guest Editor, Editorial Board member of 13 journals, including, Scientific Reports, Composites Science & Technology, etc.